**Lectures in Mathematics
ETH Zürich**
Department of Mathematics
Research Institute of Mathematics

Managing Editor:
Oscar E. Lanford

Gilbert Baumslag
Topics in Combinatorial Group Theory

1993

Birkhäuser Verlag
Basel · Boston · Berlin

Author:

Gilbert Baumslag
Department of Mathematics
City College of New York
Convent Ave at 138 St
New York, NY 10031
USA

A CIP catalogue record for this book is available from the Library of Congress,
Washington D.C., USA

Deutsche Bibliothek Cataloging-in-Publication Data
Baumslag, Gilbert:
Topics in combinatorial group theory / Gilbert Baumslag. – Basel ; Boston ; Berlin :
Birkhäuser, 1993
  (Lectures in mathematics)
  ISBN 3-7643-2921-1 (Basel ...)
  ISBN 0-8176-2921-1 (Boston)

© 1993 Birkhäuser Verlag, P.O. Box 133, CH-4010 Basel, Switzerland
Printed on acid-free paper produced of chlorine-free pulp
Printed in Germany
ISBN 3-7643-2921-1
ISBN 0-8176-2921-1

9 8 7 6 5 4 3 2 1

Contents

Preface

I gave a course on Combinatorial Group Theory at ETH, Zürich, in the Winter term of 1987/88. The notes of that course have been reproduced here, essentially without change. I have made no attempt to improve on those notes, nor have I made any real attempt to provide a complete list of references. I have, however, included some general references which should make it possible for the interested reader to obtain easy access to any one of the topics treated here.

In notes of this kind, it may happen that an idea or a theorem has inadvertently been improperly acknowledged, if at all. If indeed that is the case here, I trust that I will be forgiven in view of the informal nature of these notes.

Acknowledgements

I would like to thank R. Suter for taking notes of the course and for his many comments and corrections, C.F. Miller for pointing out a number of errors and inconsistencies, M. Triulzi for proof-reading these notes and M. Schünemann for a fine job of "TeX-ing" the manuscript.

I would also like to acknowledge the help and insightful comments of Urs Stammbach, who read each chapter carefully, making many helpful remarks and corrections.

Finally, I would like to take this opportunity to express my thanks and appreciation to him and his wife Irene, for their friendship and their hospitality over many years, and to him, in particular, for all of the work that he has done on my behalf, making it possible for me to spend many pleasurable months in Zürich.

Chapter I
History

1 Introduction

This course will be devoted to a number of topics in combinatorial group theory. I want to begin with a short historical account of the subject itself. This history, besides being of interest in its own right, will help to explain what the subject is all about.

2 The beginnings

Combinatorial group theory is a loosely defined subject, with close connections to topology and logic. Its origins can be traced back to the middle of the 19th century. With surprising frequency, problems in a wide variety of disciplines, including differential equations, automorphic functions and geometry, were distilled into explicit questions about groups. The groups involved took many forms – matrix groups, groups preserving e.g. quadratic forms, isometry groups and numerous others. The introduction of the fundamental group by Poincaré in 1895, the discovery of knot groups by Wirtinger in 1905 and the proof by Tietze in 1908 that the fundamental group of a compact finite dimensional arcwise connected manifold is finitely presented served to underline the importance of finitely presented groups. Just a short time earlier, in 1902, Burnside posed his now celebrated problem.

Problem 1 *Suppose that the group G is finitely generated and that for a fixed positive integer n*

$$x^n = 1 \quad \text{for all} \quad x \in G.$$

Is G finite?

Thus Burnside raised for the first time the idea of a finiteness condition on a group. Some 66 years later P.S. Novikov and S.I. Adian, in a remarkable series of papers, answered this question in the negative (Novikov, P.S., Adian, S.I.: *Infinite periodic groups I, II, III*, **Izv. Akad. Nauk. SSSR., Ser. Mat. 32, No. 1,2,2 (1968) 212-244, 251-524, 709-731.**)

Then, in a series of extraordinarily influential papers between 1910 and 1914, Max Dehn proposed and partly solved a number of problems about finitely presented groups, thereby heralding in the birth of a new subject, combinatorial group theory. Thus the subject came endowed and encumbered by many of the problems that had stimulated its birth. The problems were generally concerned with various classes of groups and were of the following kind: Are all the groups in a given class finite (e.g., the Burnside Problem)? Finitely generated? Finitely presented? What are the conjugates of a given element in a given group? What are the subgroups of that group? Is there an algorithm for deciding for every pair of groups in a given class whether or not they are isomorphic? And so on. The objective of combinatorial group theory is the systematic development of algebraic techniques to settle such questions. In view of the scope of the subject and the extraordinary variety of groups involved, it is not surprising that no really general theory exists. However much has been accomplished and a wide variety of techniques and methods have been developed with wide application and potential. Some of these techniques have even found a wider use, e.g., in the study of so-called free rings and their relations, in generalisations of commutative ring theory, in logic, in topology and in the theory of computing. The reader might wish to consult the book by Bruce Chandler and Wilhelm Magnus, *The History of Combinatorial Group Theory: A Case Study in the History of Ideas*, **Studies in the History of Mathematics and the Physical Sciences 9** (1982), Springer-Verlag, New York, Heidelberg, Berlin.

I do not want to stop my historical account at this point. However, in order to make it intelligible also to those who are not altogether familiar with some of the terms and notation I will use, as well as some of the theorems and definitions I will later take for granted, I want to continue my discussion, interspersing it with ingredients that I will call to mind as I need them.

3   Finitely presented groups

Let G be a group. We express the fact that H is a subgroup of G by writing $H \leq G$; if H is a normal subgroup of G we write $H \trianglelefteq G$.

Let $X \subseteq G$. Then the subgroup of G generated by X is denoted by $\mathrm{gp}(X)$. Thus, by definition, $\mathrm{gp}(X)$ is the least subgroup of G containing X. It follows that

$$\mathrm{gp}(X) = \{x_1{}^{\varepsilon_1} \ldots x_n{}^{\varepsilon_n} \mid x_i \in X, \ \varepsilon_i = \pm 1\}.$$

We call an expression of the form

$$x_1{}^{\varepsilon_1} \ldots x_n{}^{\varepsilon_n} \qquad (x_i \in X, \ \varepsilon_i = \pm 1)$$

an X-*word* or simply a *word*. Thus an X-word is simply a sequence of elements of $X \cup X^{-1}$ where $X^{-1} = \{x^{-1} \mid x \in X\}$. Notice that each X-word

$$w = x_1^{\varepsilon_1} \ldots x_n^{\varepsilon_n} \qquad (x_i \in X, \ \varepsilon_i = \pm 1)$$

takes on a value g in G after multiplying together, in G, the various $x^{\pm 1}$ that appear in w. We sometimes express this fact by writing

$$w =_G g.$$

An X-word

$$w = x_1^{\varepsilon_1} \ldots x_n^{\varepsilon_n} \qquad (x_i \in X, \ \varepsilon_i = \pm 1)$$

is termed *reduced* if

$$x_i = x_{i+1} \text{ implies } \varepsilon_i + \varepsilon_{i+1} \neq 0 \qquad (i = 1, \ldots, n-1).$$

If $G = \mathrm{gp}(X)$ and every non-empty reduced X-word $w \neq_G 1$ then we term X a *free set of generators* of G and G itself is termed *free*; we also say that X *freely generates* G or that G *is free on* X. Notice that if G is free on X, then two reduced X-words have equal values in G if and only if they are identical; this allows us to identify reduced X-words in a free group with the elements they represent, and we shall often tacitly take advantage of this observation.

The following theorem is not hard to prove; the interested reader can either supply a proof or consult one of the references cited at the end of the chapter.

Theorem 1   *(i) Given any set X there exists a free group G freely generated by X, called the free group on X.*
(ii) If G is free on X and also on Y, then $|X| = |Y|$; this common cardinal number is termed the rank of the free group G.
(iii) Let G be free on X. Then for every group H and every map $\theta : X \longrightarrow H$ there exists a unique homomorphism $\varphi : G \longrightarrow H$ such that $\varphi \mid X = \theta$.
(iv) Let G be a group and suppose that X generates G. If every reduced X-word is different from 1 in G, then G is a free group, freely generated by X.

Corollary 1   *Every group is isomorphic to a factor group of a free group.*

A group is termed an α-*generator* group if it can be generated by a set of cardinality α, it is termed *finitely generated* if it can be generated by a set of finite cardinality.

Let G again be a group, $X \subseteq G$. Then the least normal subgroup of G containing X, the so-called *normal closure* of X in G, is denoted by $\mathrm{gp}_G(X)$. So

$$\mathrm{gp}_G(X) = \mathrm{gp}(g^{-1}xg \mid g \in G, \ x \in X).$$

Now suppose G is a group, F a free group on X, θ a map from X into G such that

$$G = \mathrm{gp}(X\theta).$$

Then the extension φ of θ to F maps F onto G with kernel K. Suppose

$$K = \mathrm{gp}_F(R).$$

Then we write

$$G = \langle X; R \rangle \tag{1}$$

and term $\langle X; R \rangle$ *a presentation* of G. Notice that such a presentation (1) comes with an implicit map $\theta : X \longrightarrow G$, termed a *presentation map*, such that the extension of θ to the free group F on X yields an epimorphism φ with kernel $\mathrm{gp}_F(R)$.

If we identify X with its image in G then (1) simply means that X generates G and everything about G can be deduced from the fact that $r = 1$ in G for every $r \in R$.

Example 1 *Let*

$$G = \langle\, a, b \;;\; a^{-1}bab^{-2}, b^{-1}aba^{-2} \,\rangle.$$

Notice that in G

$$a^{-1}ba = b^2 \;,\;\; b^{-1}ab = a^2.$$

So

$$a = a^{-1}a^2 = a^{-1}b^{-1}ab = (a^{-1}ba)^{-1}b = b^{-2}b = b^{-1}.$$

But then this implies

$$b = b^2 \qquad or \qquad b = 1.$$

So

$$a = 1.$$

In other words

$$G = \{1\},$$

i.e., G is a group with exactly one element. (We refer to such groups as trivial groups, or simply as trivial.)

The lesson here is that groups given by presentations can be very tricky, even if the presentations are finite. The class of groups that have such finite presentations are termed *finitely presented* or, sometimes, *finitely presentable*. In more detail, then, we have the following

Definition 1 *A group is finitely presented if it has a finite presentation, i.e. if*

$$G = \langle X; R \rangle$$

where X and R are both finite.

Since every group is a factor group of a free group, every group has a presentation, indeed infinitely many presentations. Notice that if X is any non-empty set and R is a subset of the free group F on X, then the pair $\langle X; R \rangle$ can be thought of as a presentation of the group F/K, where $K = \mathrm{gp}_F(R)$. In this case the presentation map is the map which sends x to xK. We refer then to the pair $\langle X; R \rangle$ itself, as a presentation and we sometimes term F/K *the group of the presentation* $\langle X; R \rangle$. We term the presentation $\langle X; R \rangle$ *finite* if both X and R are finite, *finitely generated* if X is finite, and so on; the elements of R are termed *defining relators* and those of $\mathrm{gp}_F(R)$ are termed *consequences of R* or *relators*.

It is time now to return to more history and to Dehn.

4   More history

In his paper in 1912 Dehn explicitly raised three problems about finitely presented groups.

The word problem
Let G be a group given by a finite presentation

$$G = \langle X; R \rangle.$$

Is there an algorithm which decides whether or not any given $X-$word w represents the identity in G, i.e., whether or not $w =_G 1$?

The conjugacy problem
Let G be a group given by a finite presentation

$$G = \langle X; R \rangle.$$

Is there an algorithm which decides whether or not any pair of words v, w represent conjugate elements in G, i.e., if there exists an X-word z such that

$$w =_G z^{-1}vz?$$

Finally:

The isomorphism problem
Is there an algorithm which determines whether or not any pair of finite presentations (in some well-defined class of finite presentations) define isomorphic groups?

The point of working out Example 1 becomes a little clearer when viewed in the light of these problems of Dehn. In fact all three of them arose naturally in Dehn's work on the fundamental groups of two-dimensional surfaces. The question as to whether a given loop is homotopic to the identity is the word problem, whether two loops are freely homotopic is the conjugacy problem and whether the fundamental groups of two surfaces are isomorphic relates to the problem as to whether the spaces are homeomorphic.

Some 20 years after Dehn proposed these problems, Magnus proved his famous Freiheitssatz:

Theorem 2 (W. Magnus 1930) *Let G be a group with a single defining relator, i.e.,*

$$G = \langle x_1, \ldots, x_q; r \rangle.$$

Suppose r is cyclically reduced, i.e., the first and last letters in r are not inverses of each other. If each of $x_1, \ldots, x_q$ actually appears in r, then any proper subset of $\{x_1, \ldots, x_q\}$ freely generates a free group.

This led to the first major break-through on the word problem.

Theorem 3 (W. Magnus 1932) *The word problem for a 1-relator group has a positive solution.*

It took almost 50 years before all of Dehn's questions were finally answered.

First, in 1954, P.S. Novikov proved the remarkable

Theorem 4   *There exists a finitely presented group with an insoluble word problem.*

Notice that such a group with an insoluble word problem also has an insoluble conjugacy problem. Novikov's proof was a combinatorial tour-de-force. New and simpler proofs were obtained by W.W. Boone in 1959 and J.L. Britton in 1961.

The very existence of a finitely presented group with an insoluble word problem led S.I. Adian, in 1957, to prove a most striking negative theorem about finitely presented groups. In order to explain we need the notion of a *Markov property*.

Definition 2 *An algebraic property (i.e., one preserved under isomorphism) of finitely presented groups, is termed a Markov property, if*
(i)   there exists a finitely presented group with the property,
(ii) there exists a finitely presented group which cannot be embedded in, i.e. is not isomorphic to a subgroup of, a group with the property.

Here is the formulation in 1958 of Adian's theorem by M.O. Rabin:

Theorem 5 (Adian 1957, Rabin 1958) *Let $\mathcal{M}$ be a Markov property. Then there is no algorithm which decides whether or not any finite presentation defines a group with the property $\mathcal{M}$.*

To illustrate, we observe, without proof (the first three observations require none) that the following are Markov properties:

(i) *triviality;*
(ii) *finiteness;*
(iii) *commutativity;*
(iv) *having solvable word problem;*
(v) *simplicity;*
(vi) *freeness.*

The upshot of this theorem is, loosely put, that given almost any property of groups it is algorithmically undecidable as to whether or not any finite presentation defines a group with this property.

Notice that the seemingly haphazard proof of the fact that the group in Example 1 is trivial was no accident or lack of skill – the insolubility of the triviality problem makes such proofs ad hoc, by necessity.

Adian's theorem was followed in 1959 by similar, much easier, theorems by Baumslag, Boone and B.H. Neumann, about elements and subgroups of a finitely presented group.

Theorem 6   *There is a finitely presented group G_0 such that no effective procedure exists to determine whether or not a word in the generators of G_0 represents*
(i)   *an element in the center of G_0;*
(ii) *an element permutable with a given element of G_0;*
(iii) *an n-th power with $n > 1$ an integer;*
(iv) *an element whose conjugacy class is finite;*
(v)   *an element of a given subgroup of G_0;*
(vi) *a commutator, i.e. of the form $x^{-1}y^{-1}xy$;*
(vii) *an element of finite order.*

Theorem 7   *Let $\mathcal{P}$ be an algebraic property of groups. Suppose*
(i) *there is a finitely presented group that has $\mathcal{P}$;*
(ii) *there exists an integer n such that no free group of rank r has $\mathcal{P}$ if $r \geq n$.*
Then there is a finitely presented group G such that there is no algorithm which determines whether or not any finite set of elements of G generates a subgroup with $\mathcal{P}$.

So e.g., there is a finitely presented group G such that there is no algorithm which decides whether or not any finitely generated subgroup of G is finite.

It follows that, from an algorithmic standpoint, finitely presented groups consti-
tute a completely intractable class. For a general reference to this subject see the
paper by C.F. Miller III: *Decision problems for groups–survey and reflections* in
Algorithms and Classification in Combinatorial Group Theory MSRI Publications
No. 23, edited by G. Baumslag and C.F. Miller III, Springer-Verlag (1991).

5   Higman's marvellous theorem

In a sense, one aspect of the theory of finitely presented groups was brought to a
close in 1961, when Graham Higman proved the following extraordinary

Theorem 8   *Let G be a finitely generated group. Then G is a subgroup of a
finitely presented group if and only if G has a presentation of the form $\langle X; R \rangle$,
where X is finite and R is a recursively enumerable subset of the free group on X.*

It follows from this theorem of Higman (G. Higman: *Subgroups of finitely presented
groups*, **Proc. Royal Soc. London Ser. A 262**, 455-475 (1961)) that there is a close
connection between recursive function theory and the subgroup structure of finitely
presented groups.

I want to briefly illustrate just how powerful Higman's theorem is, by concocting
a finitely presented group with an unsolvable word problem.

To this end let f be a function with domain and codomain the positive integers
and suppose that

(i)   given any positive integer n we can effectively compute $f(n)$;
(ii) given any positive integer m there is no effective method which decides wheth-
      er or not there is a positive integer n such that $f(n) = m$.

So f is a recursive (or computable) function whose range is not a recursive subset
of the positive integers. Such functions do exist.

Now form

$$G = \langle \, a, b, c, d \, ; \, b^{-f(n)} a b^{f(n)} = c^{-f(n)} dc^{f(n)} \quad (n \geq 1) \rangle.$$

G is a *finitely generated, recursively presentable group*, i.e., it has a presentation on
a finite number of generators and a recursively enumerable set of defining relations.
Moreover it can be shown (see Chapter VI) that

$$b^{-m} a b^{m} =_G c^{-m} dc^{m} \text{ if and only if } m = f(n) \text{ for some positive integer } n.$$

Thus

$$w_m = b^{-m}ab^m c^{-m}d^{-1}c^m =_G 1$$

if and only if

$$m = f(n) \text{ for some positive integer } n.$$

This means that in order to solve the word problem for G we need to be able to decide whether or not any given positive integer is in the range of f. Such a decision is impossible by our choice of f. Now G is not finitely presented. But by Higman's theorem G can be embedded in a finitely presented group H. We claim that H has an unsolvable word problem, for otherwise we could apply the solution of the word problem for H to the words w_m and decide whether or not $w_m =_G 1$.

Higman's theorem leaves open the nature of an arbitrary subgroup of a finitely presented group.

6 Varieties of groups

I want next to turn my attention to an extremely interesting topic in group theory that received much attention in the 1960's. Although the subject lapsed into disfavour for a while, a lot of old and seemingly impossible open problems have been solved recently. The results are intriguing enough to merit some discussion. (See the book by Hanna Neumann: *Varieties of Groups*, **Ergebnisse der Mathematik und ihrer Grenzgebiete 37**, Springer-Verlag Berlin Heidelberg New York (1967) for an introduction to varieties.)

Let me start out with a definition.

Definition 3 *A non-empty class $\mathcal{V}$ of groups is termed a variety (of groups) if it is closed under homomorphic images, subgroups and cartesian products.*

In order to give some examples, I need to introduce some notation. Suppose that G is a group, $x, y \in G$. Then the *commutator* $x^{-1}y^{-1}xy$ of x and y is denoted by $[x, y]$ and the *conjugate* x^y of x by y is denoted by x^y. Thus

$$[x, y] = x^{-1}y^{-1}xy , \quad x^y = y^{-1}xy .$$

It is easy to check that the following identities, which we will refer to as *the basic commutator identities*, hold:

$$\begin{aligned}
x^y &= x[x, y] \\
(xy)^z &= x^z y^z \\
[x, y]^{-1} &= [y, x] \\
[xy, z] &= [x, z]^y [y, z] \\
[x, yz] &= [x, z][x, y]^z .
\end{aligned}$$

These basic identities can be verified by direct calculation.

Suppose H and K are subgroups of G. Then we define

$$[H, K] = \mathrm{gp}([h, k] \mid h \in H, \ k \in K).$$

The *commutator subgroup* or *derived group* of G is denoted by G' and is defined by

$$G' = [G, G].$$

Note that if $H \trianglelefteq G$, $K \trianglelefteq G$, then $[H, K] \trianglelefteq G$. So $G' \trianglelefteq G$ and is the smallest normal subgroup of G with an abelian factor group. Inductively we define

$$G^{(n)} = (G^{(n-1)})' \qquad (n \geq 2)$$

where $G^{(1)} = G'$, and the series

$$G = G^{(0)} \geq G' \geq G^{(2)} \geq \ldots \geq G^{(n)} \geq \ldots$$

is termed the *derived series* of G. G is termed *solvable* if $G^{(n)} = 1$ for some n, the least such n being termed the *derived length* of G. Notice that subgroups, homomorphic images and cartesian products of solvable groups of derived length at most d are again solvable of derived length at most d. Thus the class S_d of all solvable groups of derived length at most d is a variety.

Now let K be a field. Then $\mathrm{GL}(n, K)$ denotes the group of all n × n invertible matrices over K and $\mathrm{Tr}(n, K)$ the subgroup of all lower triangular matrices (i.e. zeroes above the main diagonal) of $\mathrm{GL}(n, K)$. We denote, in the case where K is commutative, the subgroup of $\mathrm{GL}(n, K)$ of matrices of determinant 1 by $\mathrm{SL}(n, K)$. Note that in the commutative case $\mathrm{Tr}(n, K)$ is solvable. In particular, if $K = \mathcal{Q}(x)$, the field of fractions of the polynomial algebra $\mathcal{Q}[x]$ in a single variable, we have the two triangular groups

$$N = \mathrm{gp}\left(a = \begin{pmatrix} 1 & 0 \\ 1 & 1 \end{pmatrix}, \ d = \begin{pmatrix} 3/2 & 0 \\ 0 & 1 \end{pmatrix} \right)$$

and

$$W = \mathrm{gp}\left(a, \ t = \begin{pmatrix} x & 0 \\ 1 & 1 \end{pmatrix} \right).$$

We leave it to the reader to prove.

Exercise 1

(a) $N' \cong Z[3/2][2/3] = Z[1/6]$ *the additive group of the ring of integers with 1/6 adjoined;*

(b) W' *is a free abelian group of infinite rank;*

(c) $N'' = W'' = 1$, *i.e., N and W are metabelian.*

It follows, e.g., from (b), that a subgroup of a finitely generated (metabelian) group need not be finitely generated.

The *centre* of G is denoted by ζG; so

$$\zeta G = \{x \in G \mid [x,y] = 1 \quad \text{for all} \quad y \in G\}.$$

The *upper central series* of G is defined to be the series

$$1 = \zeta_0 G \leq \zeta_1 G \leq \cdots \zeta_n G \leq \cdots$$

where inductively

$$\zeta_{i+1}G/\zeta_i G = \zeta(G/\zeta_i G) \qquad (i \geq 0).$$

So $\zeta_1 G = \zeta G$. The *lower central series* of G is defined to be the series

$$G = \gamma_1 G \geq \gamma_2 G \geq \cdots \geq \gamma_n G \geq \cdots$$

where inductively

$$\gamma_{n+1}G = [\gamma_n G, G] \qquad (n \geq 1).$$

G is *nilpotent* if $\gamma_{c+1}G = 1$ for some c with the least such c the *class* of G.

It follows that

$$\zeta_c G = G \quad \text{if and only if} \quad \gamma_{c+1}G = 1 .$$

Notice that subgroups, homomorphic images and cartesian products of nilpotent groups of class at most c are again nilpotent of class at most c. Thus the class $\mathcal{N}_c$ of all nilpotent groups of class at most c is a variety of groups. A group that is nilpotent of class at most c is solvable of derived length at most c.

Exercise 2 *If G is a finitely generated nilpotent group of class c, prove (by induction) that $\gamma_c G$ is finitely generated and hence that every subgroup of G is finitely generated.*

Now suppose that $\mathcal{U}$ and $\mathcal{V}$ are varieties of groups. Then we define their product $\mathcal{U}.\mathcal{V}$ to be the class of all groups G which contain a normal subgroup N such that $N \in \mathcal{U}$ and $G/N \in \mathcal{V}$. We refer to the groups in $\mathcal{U}.\mathcal{V}$ as $\mathcal{U}$ by $\mathcal{V}$ groups. It is not hard to see that the product $\mathcal{U}.\mathcal{V}$ is again a variety. It is also not hard to see that this product is associative and that it turns the set of all varieties of groups into a semigroup with an identity. In 1962 Bernhard H., Hanna and Peter M. Neumann (and independently at about the same time A.I. Smelkin) proved that this semigroup of varieties of groups is free. In more detail, let $\mathcal{E}$ denote the variety of trivial groups and let $\mathcal{O}$ denote the variety of all groups. Moreover let us term a variety $\mathcal{V}$ of groups, $\mathcal{V} \neq \mathcal{O}, \mathcal{V} \neq \mathcal{E}$ *indecomposable* if it cannot be expressed as a product of varieties different from $\mathcal{O}$, $\mathcal{E}$. Then we have the following

Theorem 9   *The set of all varieties of groups with binary operation product is a semigroup with the property that every variety different from $\mathcal{E}$ and $\mathcal{O}$ can be written as a product of indecomposable varieties in exactly one way.*

In 1970, A. Ju. Olshanskii, and, at about the same time, M.R. Vaughan-Lee and S.I. Adian (see the paper by L.G. Kovacs and M.F. Newman at the end of this chapter) proved the

Theorem 10   *There exist continuously many different varieties of groups.*

More recently Kleiman has obtained a number of remarkable negative results about varieties of groups (see the paper by C.F. Miller cited earlier for further details).

There is in each variety of groups a counterpart to the notion of a free group, defined in terms of a universal mapping property. More precisely

Definition 4 *Let $\mathcal{V}$ be a variety of groups. Then a group F in $\mathcal{V}$ is said to be free in $\mathcal{V}$ or a free $\mathcal{V}$-group, if it is generated by a set X such that for every group G in $\mathcal{V}$ and every mapping θ from X into G, there exists a homomorphism ϕ of F into G which agrees with θ on X.*

We shall need the

Definition 5 *A group G is termed hopfian if $G \cong G/N$ implies that $N=1$.*

Finitely generated free groups have this property. In 1989, somewhat surprisingly, S.V. Ivanov proved

Theorem 11 *There exists a variety $\mathcal{V}$ such that all of the non-cyclic free groups in $\mathcal{V}$ are not hopfian.*

This remarkable theorem was proved by using a variation of *small cancellation theory*. I will have a little more to say about this later. Many of the ideas involved can be traced back to Dehn, Tartakovskii, Adian and Novikov, and Olshanskii (see the book by Roger C. Lyndon and Paul E. Schupp: *Combinatorial Group Theory*, **Ergebnisse der Mathematik und ihrer Grenzgebiete 89**, Springer-Verlag, Berlin, Heidelberg, New York (1977)).

I have already mentioned Burnside's question about the existence of finitely generated infinite groups, all of whose elements are of finite order. As noted earlier, Burnside's question was answered by P.S. Novikov and S.I. Adian in 1968 who proved the following theorem (see the book by S.I. Adian: *The Burnside Problem and Identities in Groups*, **Ergebnisse der Math. und ihrer Grenzgebiete 95**, Springer-Verlag, Berlin, Heidelberg, New York, Translated from the Russian by John Lennox and James Wiegold).

Theorem 12 *There exists, for every odd integer $n > 4381$ and every integer $m > 1$ an infinite m−generator group all of whose elements have finite order dividing n.*

The bound 4381 has been improved to 665 (see the book by Adian cited above).

Proceeding in a slightly different direction, A. Ju. Olshanskii has proved, by using methods of small cancellation theory, the

Theorem 13 *For every sufficiently large prime p (e.g. $p > 10^{75}$) there exists an infinite group all of whose proper subgroups are of order p.*

The existence of an infinite group all of whose proper subgroups are finite was first raised by Tarski and, in his honour, groups with this property are now known as *Tarski monsters*.

We concentrate next on the class of solvable groups. It should be noted here that a finitely generated abelian group is a direct product of a finite number of cyclic groups. So all the algorithmic problems mentioned at the outset can be solved for them. Finitely presented solvable groups can be viewed as generalisations of these finitely generated abelian groups. In 1981, O. Kharlampovich proved the

Theorem 14 *There exists a finitely presented solvable group with an insoluble word problem.*

This, allied with the following theorem of Baumslag, Strebel and Gildenhuys (1985) puts the class of finitely presented solvable groups in much the same place as that of finitely presented groups as a whole.

Theorem 15 *The isomorphism problem for finitely presented solvable groups is recursively undecidable.*

Positive algorithmic results about finitely presented groups are few and far between. In 1980, Grunewald and Segal proved the deep

Theorem 16 *The isomorphism problem for finitely generated nilpotent groups has a positive solution.*

More recently Segal has extended this result to the larger class of *polycyclic groups*, i.e. groups which can be obtained from the trivial group by finitely many extensions by cyclic groups.

The work of Grunewald and Segal alluded to above, makes use of the theory of arithmetic and algebraic groups, and contains also a positive solution to the isomorphism problem for finite dimensional Lie algebras over $\mathbf{Q}$, a problem that had been open for almost a century.

In 1978 Bieri and Strebel introduced an invariant of finitely generated metabelian groups which detects whether or not such groups are finitely presented. Subsequently they showed that there is a similar, less discriminating, invariant of an arbitrary finitely generated group, now termed the *Bieri-Strebel invariant*. This work of Bieri and Strebel is extremely important and the interested reader is referred to the survey article by Strebel (Ralph Strebel: *Finitely Presented Soluble Groups*, in **Group Theory, Essays for Philip Hall** (1984) 257-314) for a detailed discussion of this invariant and finitely presented solvable groups as a whole (see the references at the end of the chapter).

7   Small Cancellation Theory

In his work, in 1912, on the solution of the word and conjugacy problems for surface groups, Dehn showed that, in a sense, not too much cancellation takes place on forming products of certain sets of defining relators for these groups. This point of view has led to what is now called *Small Cancellation Theory* (see the book by Lyndon and Schupp cited above for more details).

In order to explain what this means, suppose that $\langle X; R \rangle$ is a presentation of the group G:

$$G = \langle X; R \rangle \ .$$

Assume that R is *symmetrized*, i.e. closed under inverses and cyclic permutations and that all of the elements of R are cyclically reduced. We define the *length* $\ell(f)$ of an element $f \in F$ to be the number of letters in the unique reduced X-word representing f. A presentation $\langle X; R \rangle$ is called a *one-sixth presentation* if it is symmetrized and satisfies the following condition: if $r, s \in R$ and if either more than one sixth $\ell(s)$ or more than one sixth $\ell(r)$ cancels on computing the reduced X-word representing rs, then $r = s^{-1}$.

The following theorem of Greendlinger (1960) holds:

Theorem 17   *Suppose that the group G has a one-sixth presentation (as above), and that w is a non-empty reduced X-word. If $w =_G 1$, then there exists $r = a_1 \ldots a_k \in R$ such that*

$$w = b_1 \ldots b_j a_1 \ldots a_n c_1 \ldots c_l \qquad where \qquad n > \frac{k}{2} \ ,$$

i.e. w contains more than half a relator (here $b_i, a_j, c_k \in X \cup X^{-1}$).

Now suppose that the group G is given by a finite presentation $\langle X; R \rangle$, which satisfies the one-sxith condition descussed above. If w is a reduced X-word, by inspecting the finitely many elements of R, we can determine whether or not more than half of one of them is a subword of w. If not, then $w \neq_G 1$. Otherwise, $w =_G tuv$ where u is more than one half of some element $r \in R$. It follows that $r = us$, where $\ell(s) < \ell(u)/2$ and

$$w =_G ts^{-1}v.$$

Since $\ell(ts^{-1}v) < \ell(w)$, we can repeat the process with w replaced by $ts^{-1}v$. We either find, at some stage, that $w \neq_G 1$ or else that $w =_G 1$. This algorithm, known as *Dehn's algorithm*, solves the word problem for groups which are given by finite presentations satisfying the one-sixth condition. A similar, slightly more complicated argument, applies also to the conjugacy problem.

This approach to the study of groups given by generators and relators was carried further by Lyndon in 1965, who re-introduced diagrams into the study of groups allowing for the use of geometric-combinatorial arguments in handling cancellation phenomena in group theory. These methods and ideas have now been systematized and generalized, yielding important and powerful theorems in group theory. I have already mentioned the work of Olshanskii. I should also mention the work of Rips, who independently, has refined the theory and proved a number of remarkable results about groups.

Recently, M. Gromov has created a beautiful theory of what he terms *hyperbolic* groups (M. Gromov: *Hyperbolic Groups*, in **Essays on group theory**, MSRI Publications No. 8, edited by S. Gersten, Springer-Verlag (1987); see also *Group theory from a geometrical standpoint*, Edited by E. Ghys, A. Haefliger and A. Verjovsky, World Scientific (1991)). I. G. Lysenok and, independently, M. Shapiro, have proved that hyperbolic groups can be characterised as follows. A group G is *hyperbolic* if and only if it has a finite presentation

$$G = \langle X; R \rangle$$

with the following property: if w is a reduced X-word and $w =_G 1$, then w contains more than one half of a defining relator. So the word problem for hyperbolic groups can be solved by the kind of algorithm that Dehn introduced eighty years ago. The class of hyperbolic groups is contained in a somewhat wider class of groups called *automatic groups*. This latter class were introduced by Cannon, Epstein, Holt, Paterson and Thurston (J.W. Cannon, D.B.A. Epstein, D.F. Holt, M.S. Paterson and W.P. Thurston: *Word processing and group theory*, preprint, University of Warwick (1991)). They are of great current interest, because of their connection with hyperbolic geometry, automata and computing.

There are some other important developments that I want to mention. These include the *Bass-Serre theory of groups acting on trees* (see Chapter VII) and the on-going study of the cohomology of groups. Here we have the remarkable theorem of J.R. Stallings and R. Swan:

Theorem 18   *A group of cohomological dimension one is free.*

This was proved first by Stallings for countable groups, with the resulting consequence

Corollary 1 *A torsion-free group with a free subgroup of finite index is free.*

The monograph by J.R. Stallings: *Group Theory and 3-dimensional Manifolds,* **Yale Monographs 4** (1971), is an excellent reference for this topic.

In addition I want also to mention the work of J. McCool on finitely presented subgroups of the automorphism groups of finitely generated free groups and the graph-theoretic methods of Stallings with applications by Gersten to fixed point subgroups of automorphisms of free groups. In particular Gersten has proved the following theorem (S.M. Gersten: *On Fixed Points of Certain Automorphisms of Free Groups,* **Proc. London Math. Soc. 48** (1984), 72-94).

Theorem 19   *Let F be a finitely generated free group and let φ be an automorphism of F. Then*
$$\operatorname{Fix}\varphi = \{a \in F \mid a\varphi = a\}$$
is finitely generated.

The following additional references may be useful to the interested reader.

Kovacs, L.G. and M.F. Newman, *Hanna Neumann's Problems on Varieties of Groups,* **Proc. Second Internat. Conf. Theory of Groups Canberra** (1973), 417-433.

Kurosh, A.G., *The theory of groups, 2nd edition,* **translated from the Russian by K.A. Hirsch, vols. I and II,** Chelsea Publishing Company, New York (1955).

Magnus, Wilhelm, Abraham Karrass and Donald Solitar, *Combinatorial Group Theory,* Dover Publications, Inc., New York (1976).

The book cited earlier by Lyndon and Schupp is a good source of information for many of the topics discussed in this chapter.

Chapter II
The Weak Burnside Problem

1 Introduction

In 1902 Burnside wrote "A still undecided problem in the theory of discontinuous groups is whether the order of a group may be not finite while the order of every operation it contains is finite". He tacitly assumed that the groups involved are all finitely generated.

In fact this quotation of Burnside has now been turned into the so-called Burnside Problem, which I formulated in Chapter I.

The Burnside Problem
Let G be a finitely generated group. If for some fixed positive integer n

$$x^n = 1 \quad for \ all \quad x \in G,$$

is G finite?

Burnside proved that the answer is in the affirmative for $n = 2, 3$.

Exercise 1 *Prove that Burnside's Problem has an affirmative answer when $n = 2$ or for arbitrary n when the group is abelian.*

In 1940 Sanov settled Burnside's Problem positively for $n = 4$. The answer is again positive when $n = 6$; this is due to Marshall Hall, Jr. (see Marshall Hall, Jr.: *The theory of groups*, **Chelsea Publishing Company** New York, N.Y. (1976)) for detailed references). The case $n = 5$ is unresolved.

As we have already noted in Chapter 1, Burnside's Problem was solved, negatively, by Novikov and Adian in 1968.

There are other forms of this problem.

The Weak Burnside Problem

Let G be a finitely generated group. Suppose that every element of G is of finite order. Is G finite?

If V is a finite dimensional vector space, then we denote by $GL(V)$ the group of all invertible linear transformations of V. Burnside himself solved the Weak Burnside Problem for the finitely generated subgroups of $GL(V)$, when the ground field is the field of complex numbers.

In 1964, E.S. Golod showed that the answer to the Weak Burnside Problem is in the negative (E.S. Golod: *Nil-algebras and residually finite groups*, **Izv. Akad. Nauk. SSSR Ser. Mat. 28, No.2 (1964), 273-276** (see also the reference, cited below, to Golod and Shafarevich).

Theorem 1 *There exists, for each prime p, a 2-generator infinite group all of whose elements are of finite order a power of the prime p.*

There is one other facet of the Burnside Problem that has attracted much attention, partly because of its connection with the theory of Lie rings.

The Restricted Burnside Problem

Let r and n be fixed positive integers. Is there a bound on the orders of the finite groups G with r generators satisfying the condition

$$x^n = 1 \quad for \ all \quad x \in G \ ?$$

Here the major result is due to Kostrikin:

Theorem 2 *If n is any prime, then the Restricted Burnside Problem has a positive answer.*

This has very recently been extended by Zelmanov to the case where n is an arbitrary power of a prime.

My primary objective in this chapter is to give a negative solution to the Weak Burnside Problem. The basic idea is due to Grigorchuk, although the point of view and the exposition I shall give here is due to Gupta and Sidki. Because of this I have elected to call the groups described here the Grigorchuk-Gupta-Sidki groups.

2   The Grigorchuk-Gupta-Sidki groups

Let me recall that a group is called a p-group, where here p is a prime, if every element is of order a power of p. Then we have the following

Theorem 3 (Grigorchuk, Gupta & Sidki) *There exists, for every odd prime p, a tree X whose automorphism group contains a 2-generator, infinite p-group G.*

We will restrict attention to the case where $p = 3$. The Grigorchuk-Gupta-Sidki group G is defined as a subgroup of the group of automorphisms of the tree X of Fig.(II,1)

Each vertex v of X is the base of another tree $X(v)$ isomorphic to X. Thus

$$X = X(\bullet) \ .$$

Using this notation we can redraw X as follows in Fig.(II,2).

Notice the labelling system: if v is any given vertex, other than the initial vertex $\bullet$, then the other vertex of the left-most edge emanating from v is labelled $v1$, the middle vertex is labelled $v2$ and the right-most vertex is labelled $v3$. Observe that if you stand at any vertex in this graph and look upwards you have exactly the same view.

The Grigorchuk-Gupta-Sidki group is generated by two elements τ and α which we define by specifying their action on the vertices of X and deducing what happens to the edges. Notice that since

$$X = X(\bullet) \ \cong \ X(v)$$

for every vertex v, each automorphism β of X has associated to it a corresponding automorphism of $X(v)$ which we denote by $\beta(v)$. This is not to be confused with the image of v under β which, using algebraic notation, is properly denoted $v\beta$.

First the definition of τ:

τ is the automorphism of X which permutes $X(1)$, $X(2)$ and $X(3)$ cyclically, mapping $X(1)$ naturally onto its isomorphic image $X(2)$, $X(2)$ onto $X(3)$, $X(3)$ onto $X(1)$ and leaving the base $\bullet$ of X fixed.

We will find it convenient also, given an automorphism γ of $X(v)$ which leaves the vertex v fixed, to continue γ to an automorphism of X. We define the extension of γ to X by making it leave everything outside $X(v)$ fixed. This automorphism of X we again denote by γ.

Next the definition of α:

$$\alpha = \tau(1)\ \tau(2)^{-1}\alpha(3)\ . \tag{1}$$

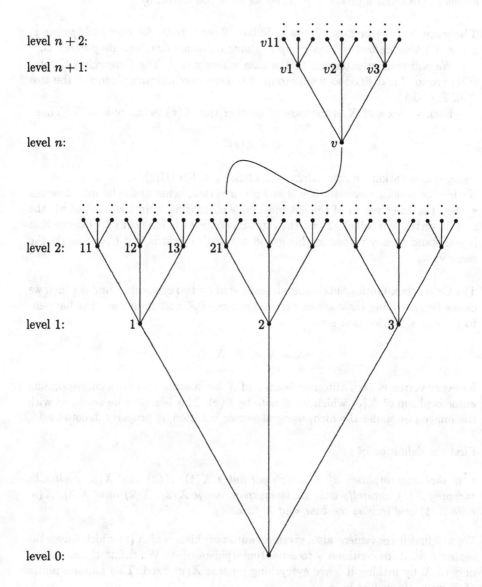

level $n+2$:

level $n+1$:

level n:

level 2:

level 1:

level 0:

Fig.(II,1) Graph X

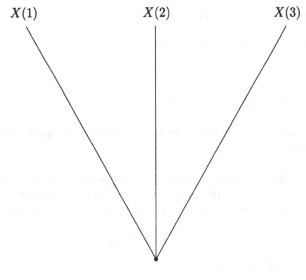

$$X(1) \qquad\qquad X(2) \qquad\qquad X(3)$$

Fig.(II,2)

The definition of α, although awkward, is unambiguous if it is suitably interpreted. On $X(1)$ α is simply $\tau(1)$, on $X(2)$ it is $\tau(2)^{-1}$ and on $X(3)$ it is $\alpha(3)$. Each of these automorphisms is to be viewed as an automorphism of X, in the manner specified above, and hence leaves the vertex $\bullet$ fixed. Thus the definition (1) implies that α leaves $\bullet$ fixed. In general, we interpret (1) to imply that for each vertex v

$$\alpha(v) = \tau(v1)\,\tau(v2)^{-1}\alpha(v3)\ ,$$

which means that α is completely determined by the automorphisms $\tau(v)$. More precisely, notice that if v is a vertex at level n, then

$$
\begin{aligned}
\alpha &= \tau(1)\,\tau(2)^{-1}\,\alpha(3)\\
&= \tau(1)\,\tau(2)^{-1}\,\tau(31)\,\tau(32)^{-1}\alpha(33)\\
&= \ldots\\
&= \tau(1)\,\tau(2)^{-1}\,\tau(31)\,\tau(32)^{-1}\ldots\,\tau(\underbrace{3\ldots3}_{n-1}1)\tau(\underbrace{3\ldots3}_{n-1}2)^{-1}\alpha(\underbrace{3\ldots3}_{n}). \qquad (2)
\end{aligned}
$$

It follows that the effect of α on v is either to leave v fixed or else is obtained by applying the appropriate $\tau(*)$. Consequently, α leaves, at each level, the right-most vertex fixed.

Our objective now is to prove that the group

$$G = \mathrm{gp}(\alpha, \tau)$$

has the desired properties.

Lemma 1 α and τ are of order 3.

Proof Clearly $\alpha \neq 1 \neq \tau$. We note first that

$$\tau^3 = 1 .$$

This follows from the very definition of τ.

Next we prove that $\alpha^3 = 1$. It is enough to check that α^3 leaves every vertex fixed. This is clear from (2).

But let's give a slightly different argument. Suppose v is a vertex at level n. If $n = 0$, α leaves v fixed and therefore so does α^3. Inductively let us assume α^3 leaves every vertex at level $\leq n - 1$ fixed and that $n > 0$. If v is a vertex in $X(1)$ then

$$v \, \alpha^3 = v \, \tau(1)^3 = v$$

and if v is a vertex in $X(2)$,

$$v \, \alpha^3 = v \, \tau(2)^{-3} = v .$$

Finally if v is a vertex in $X(3)$,

$$v \, \alpha^3 = v \, \alpha(3)^3 .$$

But $X(3) \cong X$ and v is at level $n - 1$ in $X(3)$. Hence the inductive assumption yields

$$v \, \alpha(3)^3 = v$$

as required.   ∎

Lemma 2 G is an infinite group.

Proof Let v be a vertex in X. We put

$$H(v) = \mathrm{gp}\Big(\, \alpha(v) \, , \, \tau(v)^{-1}\alpha(v)\tau(v) \, , \, \tau(v)^{-2}\alpha(v)\tau(v)^2 \, \Big)$$

and

$$H = H(\bullet) = \mathrm{gp}\big(\, \alpha \, , \, \tau^{-1}\alpha\tau \, , \, \tau^{-2}\alpha\tau^2 \, \big).$$

Now

$$H \trianglelefteq G .   \tag{3}$$

This is clear since H is generated by all of the conjugates of α under the powers of τ.

Next we compute the forms of these conjugates of α. First

$$\alpha = \tau(1)\tau(2)^{-1}\alpha(3) . \tag{4}$$

It follows that

$$\tau^{-1}\alpha\tau = \alpha(1)\tau(2)\tau(3)^{-1} \tag{5}$$

and

$$\tau^{-2}\alpha\tau^2 = \tau(1)^{-1}\alpha(2)\tau(3) . \tag{6}$$

These assertions can be checked by simply carrying out, e.g., τ^{-1}, α and τ in this order to get (5).

Put

$$G(v) = \mathrm{gp}\big(\ \alpha(v)\ ,\ \tau(v)\ \big)$$

where v is a vertex in X. Then $G = G(\bullet) \cong G(v)$.

Next notice that

$$\tau(1),\alpha(1)\ \in\ G(1)\quad,\quad \tau(2),\alpha(2)\ \in\ G(2)\quad,\quad \tau(3),\alpha(3)\ \in\ G(3)$$

and that

$$\mathrm{gp}\big(\ G(1),G(2),G(3)\ \big) = G(1)\times G(2)\times G(3) .$$

In particular we also note that

$$H\ \leq\ G(1)\times G(2)\times G(3) . \tag{7}$$

It follows that $\tau \notin H$, because by (7), every element of H leaves the vertices at level 1 fixed. So

$$|G/H| = 3 . \tag{8}$$

Now it follows from (4) and (5) that the projection π of H into $G(1)$ is actually onto. So if K is the kernel of π,

$$H/K\ \cong\ G .$$

Consequently we have a series

$$
\begin{matrix}
G\bullet \\
\ \ \ \Big\downarrow\ \ |G/H| = 3 \\
H\bullet \\
\ \ \ \Big\downarrow\ \ H/K \cong G \\
K\bullet
\end{matrix}
\tag{9}
$$

Thus G has a proper subgroup that maps onto G. This implies that G is infinite. ∎

We turn now to the final step in the proof.

Lemma 3 *Every element of G is of order a power of 3.*

Proof We put $a_i = \tau^{-i}\alpha\tau^i$ $(i = 0, 1, 2)$. Then by its very definition

$$H = \mathrm{gp}(a_0, a_1, a_2) \ .$$

Notice that if $h \in H$ then h can be expressed as a Y-product where $Y = \{a_0, a_1, a_2\}$. Since $a_i{}^{-1} = a_i{}^2$ we can express every such h as a *positive Y-product*, i.e., no negative powers occur:

$$h = y_1 y_2 \ldots y_n \qquad (y_i \in Y) \ . \tag{10}$$

Now an arbitrary element $g \in G$ can be written in the form (see (9))

$$g = h\tau^k \qquad (h \in H, \ 0 \le k \le 2) \ .$$

Suppose that, in addition we now write h in the positive form (10); then

$$g = y_1 y_2 \ldots y_n \tau^k \qquad (y_i \in Y, \ 0 \le k \le 2) \ . \tag{11}$$

We term this representation of g a *special representation* and define the length $l(g)$ of such a special representation of g by

$$l(g) = \begin{cases} n & \text{if } k = 0 \\ n+1 & \text{if } k \ne 0 \end{cases} \tag{12}$$

Notice that $y_1 y_2 \ldots y_n$ is a positive product involving only a_0, a_1, a_2 . We express this dependence on a_0, a_1, a_2 by using functional notation:

$$h = h(a_0, a_1, a_2) \ .$$

So

$$g = h(a_0, a_1, a_2)\tau^k \qquad (0 \le k \le 2) \ . \tag{13}$$

We want to prove that g is of order a power of 3. We do this by induction on $l(g)$.

If $l(g) \le 1$ we have already proved that g is of order dividing 3.

Suppose that we have proved that every element of G with a special representation of length at most n is of order a power of 3 and that

$$l(g) = n + 1 \qquad (n \ge 1) \tag{14}$$

with g given by (13).

We consider now two cases.

Case 1 $g = h(a_0, a_1, a_2)\tau^k$ $(0 < k \leq 2)$.

Thus here $l(h) = n$. Suppose that the number of occurrences of a_0 in $h(a_0, a_1, a_2)$ is i_0, that the number of occurrences of a_1 in $h(a_0, a_1, a_2)$ is i_1 and that the number of occurrences of a_2 in $h(a_0, a_1, a_2)$ is i_2. This means that

$$n = i_0 + i_1 + i_2 .$$

The trick is to compute

$$\begin{aligned} g^3 &= h\tau^k h\tau^k h\tau^k \\ &= h\tau^{3k}\tau^{-2k}h\tau^{2k}\tau^{-k}h\tau^k \\ &= h\tau^{-2k}h\tau^{2k}\tau^{-k}h\tau^k . \end{aligned}$$

Let's consider the cases $k = 1, 2$ in turn, starting with $k = 1$. In this instance

$$g^3 = h(a_0, a_1, a_2)\, h(a_2, a_0, a_1)\, h(a_1, a_2, a_0) . \tag{15}$$

This means that the number of occurrences of a_0 in $h(a_0, a_1, a_2)$ is i_0, the number of occurrences of a_0 in $h(a_2, a_0, a_1)$ is i_1 and the number of occurrences of a_0 in $h(a_1, a_2, a_0)$ is i_2, i.e. a_0 occurs

$$n = i_0 + i_1 + i_2$$

times in the right-hand-side of (15). A similar argument shows that each of a_1 and a_2 also occurs n times in the right-hand-side of (15) This holds true, by a similar argument, also when $k = 2$.

Now, by (4), (5) and (6),

$$\begin{aligned} a_0 &= \tau(1)\,\tau(2)^{-1}\,\alpha(3) \\ a_1 &= \alpha(1)\,\tau(2)\,\tau(3)^{-1} \\ a_2 &= \tau(1)^{-1}\,\alpha(2)\,\tau(3) . \end{aligned}$$

We can therefore re-express g^3 as an element in

$$G(1) \times G(2) \times G(3)$$

(see (7)). For definiteness let's consider the case $k = 1$.

Notice that the first component of g^3 is then

$$h\big(\tau(1),\alpha(1),\tau(1)^{-1}\big)\, h\big(\tau(1)^{-1},\tau(1),\alpha(1)\big)\, h\big(\alpha(1),\tau(1)^{-1},\tau(1)\big)\ . \tag{16}$$

It follows, from the remarks above, that $\tau(1)$ occurs n times in (16), that $\alpha(1)$ occurs n times in (16) and finally that $\tau(1)^{-1}$ also occurs n times in (16). We think of the first component of g^3 as an element of $G(1)$. As such it has a form corresponding to that of g, given by (13), i.e., we can move all the occurences of $\tau(1)$ to the right-most-side and write it in the form

$$w_1\Big(\alpha(1),\tau(1)^{-1}\alpha(1)\tau(1),\tau(1)^{-2}\alpha(1)\tau(1)^2\Big)\tau(1)^m = w_1\big(a_0(1),a_1(1),a_2(1)\big)\tau(1)^m,$$

where $a_0(1) = \alpha(1)$, $a_1(1) = \tau(1)^{-1}\alpha(1)\tau(1)$ and $a_2(1) = \tau(1)^{-2}\alpha(1)\tau(1)^2$.

Notice that this form $w_1\big(a_0(1),\, a_1(1),\, a_2(1)\big)$ is a positive word in $a_0(1)$, $a_1(1)$, $a_2(1)$ of length exactly n. And since both $\tau(1)$ and $\tau(1)^{-1}$ occur n times in (16),

$$m = 0$$

The same argument applies to each one of the components of g^3, yielding

$$g^3 = w_1 w_2 w_3 \in G(1) \times G(2) \times G(3)$$

with

$$l(w_1) = l(w_2) = l(w_3) = n\ .$$

Since $G \cong G(i)$ $(i = 1, 2, 3)$ it follows, by induction, that each of w_1, w_2, w_3 is of order a power of 3. So g is also of order a power of 3.

Case 2 $g = h(\, a_0, a_1, a_2\,)$

Here $l(g) = n+1$, i.e., $l(h) = n+1$. Now again we assume that there are i_0 occurrences of a_0 in h, i_1 occurrences of a_1 in h, and i_2 occurrences of a_2 in h. So, in this case,

$$i_0 + i_1 + i_2 = n+1\ .$$

Notice that if at least two of i_0, i_1, i_2 are zero, then h is a power of one of a_0, a_1, a_2 and hence is of order a power of 3. Therefore we may assume that at least two of i_0, i_1, i_2 are non-zero. Now view h as an element of $G(1) \times G(2) \times G(3)$; as usual:

$$\begin{aligned}
h &= h(a_0, a_1, a_2)\\
&= h\big(\tau(1),\alpha(1),\tau(1)^{-1}\big)h\big(\tau(2)^{-1},\tau(2),\alpha(2)\big)h\big(\alpha(3),\tau(3)^{-1},\tau(3)\big)\\
&= h_1\, h_2\, h_3\ .
\end{aligned}$$

On re-expressing each of these components in its special form in $G(1), G(2)$ and $G(3)$, we either find that

$$h_i = h_i' \tau(i)^k \qquad (0 < k \le 2),$$

with $l(h_i') \le n$, or else

$$h_i = h_i',$$

where again $l(h_i') \le n$. In the first case we can repeat the argument of Case 1, remembering that $G(i) \cong G$, computing $h_i{}^3$ and deducing inductively, as before, that h_i is of order a power of 3. In the second case the inductive assumption applies immediately and so again it follows that h_i is of order a power of 3. This completes the proof. ∎

The argument given above is contained in:

N. Gupta, S. Sidki: *On the Burnside Problem for Periodic Groups* **Math.Z. 182** (1983), 385-388.

As I remarked earlier, it is based on the paper

R.I. Grigorchuk: *On the Burnside Problem for Periodic Groups* English Translation: **Functional Anal. Appl. 14** (1980), 41-43.

3 An application to associative algebras

Kurosh has posed a problem for associative algebras that is analogous to that of Burnside.

Kurosh's Problem
Let A be an associative algebra over a field k. Suppose that each element a in A is the root of some polynomial $c_0 + c_1 x + \ldots + c_n x^n$ $(c_n \ne 0, n > 0)$ i.e.

$$c_0 + c_1 a + \ldots + c_n a^n = 0 .$$

Is every finitely generated subalgebra of A finite dimensional?

The first counter-example was obtained by

E.S. Golod and I.R. Shafarevich: *On towers of class fields* **Izv. Akad. Nauk SSSR Ser.Mat 28** (1964), 261-272.

The Grigorchuk-Gupta-Sidki group provides another example. In order to see how this comes about, form the group algebra $\mathbf{F}_3[G]$, where $\mathbf{F}_3$ denotes the field of three elements. So

$$\mathbf{F}_3[G] = \left\{ \sum_{\text{finite}} c_i g_i \mid c_i \in \mathbf{F}_3, g_i \in G \right\}$$

with the obvious definition of equality, coordinate-wise addition and multiplication defined by distributivity and

$$c_i g_i \cdot c_j g_j = c_i c_j (g_i g_j) \ .$$

We claim that the augmentation ideal

$$A = I(G) = \left\{ \sum c_i g_i \mid \sum c_i = 0 \right\}$$

of $\mathbf{F}_3[G]$ provides a suitable example. Indeed note first that

$$\left(\sum c_i g_i \right)^{3^m} = \sum c_i^{3^m} g_i^{3^m} = \sum c_i g_i^{3^m} \ .$$

Since each g_i is of order a power of 3, by choosing m sufficiently large we can ensure that each $g_i^{3^m} = 1$. So

$$\left(\sum c_i g_i \right)^{3^m} = \sum c_i = 0 \ .$$

It suffices then to prove that A is finitely generated since it is clearly infinite dimensional. But we claim that A is generated by

$$\alpha - 1 \, , \, \tau - 1 \ .$$

This follows from the identity

$$xy - 1 = (x-1)(y-1) + (x-1) + (y-1) \ .$$

Finally in closing, it is not hard to see that the Grigorchuk-Gupta-Sidki group G has a recursive presentation. I have been told that G is not finitely presented, but I do not know of a proof. Perhaps one way of proving this is to show that the second integral homology group of G (see Chapter VI for a definition of $H_2(G, \mathbf{Z})$) is not finitely generated. This would be of independent interest.

Chapter III
Free groups, the calculus of presentations and the method of Reidemeister and Schreier

1 Frobenius' representation

A homomorphism

$$\varphi : G \longrightarrow H$$

of a group G into another group H is termed a *representation of G*. If H is a group of permutations, then φ is termed a *permutation representation*; if H is a group of matrices over a commutative field, then φ is termed a *matrix* or *linear* representation (of G). We term a representation *faithful* if it is one-to-one. We say that a group G *acts on a set Y* if it comes equipped with a homomorphism

$$\varphi : G \longrightarrow S_Y,$$

where S_Y is the symmetric group on Y, i.e., the group of all permutations of Y. If $Y = \{1, 2, \ldots, n\}$ we sometimes write S_n in place of S_Y.

The most familiar faithful permutation representation goes back to Cayley.

Theorem 1 (Cayley) *Let G be a group. Then the map*

$$\varrho : G \longrightarrow S_G$$

defined by

$$g \longmapsto (g\varrho : x \longmapsto xg)$$

is a faithful permutation representation of G, called the right regular representation.

The proof is an immediate application of the definition.

Next I want to describe Frobenius' version of Cayley's representation. In order to do so, we need some definitions.

Definition 1   *Let G be a group, $H \le G$. Then a complete set of representatives of the right cosets Hg of H in G is a set R consisting of one element from each coset. The element in R coming from the coset Hg is denoted by $\overline{g}$ and is termed the representative of Hg or sometimes the representative of g. If, in addition, $1 \in R$, R is termed a right transversal of H in G.*

Let us put $\delta(r,g) = rg\,(\overline{rg})^{-1}$ $(r \in R, g \in G)$. Then it is easy to deduce the following lemma from the definitions.

Lemma 1
(i) $Hg = H\overline{g}$ $(g \in G)$
(ii) $\delta(r,g) \in H$
(iii) $\overline{\overline{g_1}g_2} = \overline{g_1 g_2}$ $(g_1, g_2 \in G)$
(iv) $g = \delta(1,g)\overline{g}$ $(g \in G)$

The following theorem of Frobenius is a simple consequence of Lemma 1.

Theorem 2   *Let G be a group, $H \le G$, R a complete set of representatives of the right cosets of H in G. Then the homomorphism*

$$\gamma : G \longrightarrow S_R$$

defined by

$$g \longmapsto (g\gamma : r \longmapsto \overline{rg})$$

is a representation of G, sometimes termed a coset representation of G.

This representation has turned out to be extremely useful. It leads one to some very fertile areas of investigation, *induced representation theory* and the theory of *wreath products*. I won't go into any of the details here, but extract two facts which will be useful later.

First some more notation. Suppose G acts on a set Y where φ is the ambient representation. If $y \in Y$, then the *stabilizer of y*, denoted $stab_\varphi(y)$, is by definition,

$$stab_\varphi(y) = \{\, g \in G \mid y(g\varphi) = y \,\} \le G \,.$$

We have the following theorem of M. Hall.

Theorem 3   *Let G be a finitely generated group. Then the number of subgroups of G of a given finite index j is finite.*

Proof   We concoct for each subgroup H of index j in G a homomorphism

$$\varphi_H : G \longrightarrow S_j$$

in such a way that

$$stab_{\varphi_H}(1) = H .$$

We need to define φ_H. To this end let us choose a complete set R of representatives of the right cosets of H in G. We choose an enumeration of the elements of R which is arbitrary except that the first element is in H:

$$R = \{r_1, r_2, \ldots, r_j\} \quad \text{with} \quad r_1 \in H .$$

Let γ be Frobenius' coset representation and define

$$\varphi_H : G \longrightarrow S_j$$

by

$$i(g\varphi_H) = k \qquad \Longleftrightarrow \qquad r_i(g\gamma) = r_k .$$

Notice that

$$stab_{\varphi_H}(1) = H .$$

Thus if K is a second subgroup of G of index j,

$$\varphi_H = \varphi_K \quad \Longrightarrow \quad H = K .$$

This means that the number of subgroups of G of index j is at most the number of homomorphisms of G into S_j. Now if G is an n-generator group, the number of such homomorphisms is at most

$$(j!)^n < \infty .$$

This completes the proof.   ■

We now reconstitute Cayley's representation ϱ from Frobenius' representation γ. To this end let G be a group, $H \leq G$, R a complete set of representatives of the right cosets of H in G. Think of G as being partitioned into $|R|$ disjoint blocks with $|H|$ elements in each block:

$$G = \bigcup_{r \in R} Hr \cong \bigcup_{r \in R} H \times \{r\} = H \times R .$$

If $g \in G$ then $g\varrho$ acts as follows:

$$hr(g\varrho) = hrg = h\, rg\, (\overline{rg})^{-1}\, \overline{rg} = h\, \delta(r, g)\, (r(g\gamma)) .$$

This allows us to view g as acting on the blocks $H \times \{r\}$ by first mapping each of the elements (h, r) to $(h\delta(r, g), r)$ and then bodily moving the whole block $H \times \{r\}$ to $H \times \{r(g\gamma)\}$.

There is another way of interpreting this discussion. First the functions $\delta(*, g)$ are functions from R to H. Thus let's form

$$H^R = \{ f : R \longrightarrow H \} .$$

H^R then can be thought of as a group of permutations of $H \times R$:

$$(h, r)f = \big(hf(r), r \big) .$$

And for each $g \in G$, $g\gamma$ can be thought of as a permutation of $H \times R$:

$$(h, r)(g\gamma) = \big(h, r(g\gamma) \big) .$$

Here is one consequence of this approach:

Theorem 4   *Let G be a group generated by a set X, let H be a subgroup of G and let R be a right transversal of H in G. Then*

$$H = \mathrm{gp}\big(\delta(r, x) = rx(\overline{rx})^{-1} \mid r \in R, x \in X \big) .$$

Proof   We simply trace out, for each $h \in H$, the effect of $h\varrho$ on $(1, 1) \in H \times R$. First notice that

$$(h, r)(g\varrho) = \big(h\delta(r, g), \overline{rg} \big) .$$

So, if $h \in H$,

$$(1, 1)(h\varrho) = \big(\delta(1, h), \overline{h} \big) = (h, 1) .$$

Now express h as an X-word:

$$h =_F x_1^{\varepsilon_1} \ldots x_n^{\varepsilon_n} \qquad (x_i \in X, \varepsilon_i = \pm 1) .$$

Then

$$(1, 1)(h\varrho) = (1, 1)(x_1^{\varepsilon_1}\varrho) \ldots (x_n^{\varepsilon_n}\varrho)$$
$$= \big(\delta(r_1, x_1^{\varepsilon_1}) \ldots \delta(r_n, x_n^{\varepsilon_n}), 1 \big)$$

where the r_i are the elements of R that arise from this computation. This proves that

$$h = \delta(r_1, x_1^{\varepsilon_1}) \ldots \delta(r_n, x_n^{\varepsilon_n}) .$$

Hence

$$H = \mathrm{gp}\big(\delta(r, x^{\pm 1}) \mid r \in R, x \in X \big) .$$

But

$$rx^{-1}\left(\overline{rx^{-1}}\right)^{-1} = \left(\overline{rx^{-1}x}\left(\overline{rx^{-1}x}\right)^{-1}\right)^{-1}.$$

This completes the proof. ∎

Corollary 1 *A subgroup of finite index in a finitely generated group is finitely generated.*

Exercise 1 *Prove that the intersection of two subgroups of finite index in any group is again of finite index.*

2   Semidirect products

Let

$$1 \longrightarrow A \xrightarrow{\alpha} E \xrightarrow{\beta} Q \longrightarrow 1 \tag{1}$$

be a short exact sequence of groups. We term E an extension of A by Q. If

$$1 \longrightarrow A' \xrightarrow{\alpha'} E' \xrightarrow{\beta'} Q' \longrightarrow 1$$

is another short exact sequence, we term the sequences *equivalent* if there are isomorphisms, as shown, which make the following diagram commutative:

$$
\begin{array}{ccccccccc}
1 & \longrightarrow & A & \xrightarrow{\alpha} & E & \xrightarrow{\beta} & Q & \longrightarrow & 1 \\
& & \alpha^* \downarrow \wr & & \varepsilon^* \downarrow \wr & & \xi^* \downarrow \wr & & \\
1 & \longrightarrow & A' & \xrightarrow{\alpha'} & E' & \xrightarrow{\beta'} & Q' & \longrightarrow & 1 \ .
\end{array}
$$

Every short exact sequence (1) is equivalent to the sequence

$$1 \longrightarrow A\alpha \hookrightarrow E \longrightarrow E/A\alpha \longrightarrow 1 \ .$$

We will move freely between equivalent sequences.

The sequence (1) *splits* if there exists a homomorphism

$$\eta : Q \longrightarrow E$$

such that $\eta\beta = 1$. In this case we term (1) a *split* or *splitting extension* of A by Q. Every such splitting extension (1) is equivalent to an extension

$$1 \longrightarrow \overline{A} \xrightarrow{\overline{\alpha}} \overline{E} \underset{\overline{\eta}}{\overset{\overline{\beta}}{\rightleftarrows}} \overline{Q} \longrightarrow 1$$

where
(i) $\overline{E} = \overline{Q}\,\overline{A}$, (ii) $\overline{A} \trianglelefteq \overline{E}$, (iii) $\overline{Q} \cap \overline{A} = 1$.

For simplicity of notation we simply omit all the bars. Then it follows that every element $e \in E$ can be written uniquely in the form

$$e = q\,a \qquad (q \in Q, a \in A)$$

with multiplication

$$e\,e' = q\,a\,q'a' = q\,q'a^{q'}a' \qquad (q, q' \in Q,\ a, a' \in A) . \tag{2}$$

Observe that for each $q' \in Q$, the map

$$a \longmapsto a^{q'} \qquad (a \in A)$$

is an automorphism $q'\varphi$ of A. The underlying map

$$\varphi : Q \longrightarrow \mathrm{Aut}\ A$$

is a homomorphism of Q into $\mathrm{Aut}\ A$ and the extension (1) can be reconstituted from A, Q and φ.

Conversely, let A and Q be arbitrary groups,

$$\varphi : Q \longrightarrow \mathrm{Aut}\ A$$

a homomorphism. Then we can form a split extension E of A by Q as follows. Set-theoretically

$$E = Q \times A = \{ (q,a) \mid q \in Q,\ a \in A \} .$$

We define a multiplication on E by analogy with (2):

$$(q,a)(q',a') = (qq' ,\ a(q'\varphi)a') . \tag{3}$$

It is easy to see then that E is an extension of A by Q, called the *semidirect product of A by Q* and denoted by

$$E = A \rtimes Q = A \rtimes_\varphi Q,$$

the latter to express the dependence on φ. If we identify $q \in Q$ with $(q,1)$, $a \in A$ with $(1,a)$ then we find that

(i) $E = Q\,A$,
(ii) $A \trianglelefteq E$,
(iii) $Q \cap A = 1$,

and the multiplication in E takes the form

$$q\,a\,q'a' = q\,q'\,a\,(q'\varphi)\,a' \qquad (q,q' \in Q\,,\, a,a' \in A)\,. \qquad (4)$$

We often switch from the notation (3) to the notation (4).

Examples 1 *(1) Let A be an abelian group, $Q = \langle q; q^2 = 1 \rangle$ a group of order 2. Let*

$$\varphi : Q \longrightarrow \text{Aut } A$$

be defined by

$$q\varphi : a \longmapsto a^{-1}\,.$$

Then we can form

$$E = A \rtimes Q\,.$$

(i)   If A is cyclic of order 4, E is the dihedral group of order 8.
(ii) If $A = C_{2^\infty}$ the quasicyclic group of type 2^∞, i.e., the group of all 2^n-th roots of 1 in the complex field $(n = 1, 2, \ldots)$, E is an infinite 2-group all of whose proper subgroups are either cyclic, dihedral groups or C_{2^∞}.
(iii) Compute the upper central series of the group E in (ii), above.

(2)   Let A be a free abelian group of infinite rank on

$$\ldots,\, a_{-1}\,,\, a_0\,,\, a_1\,,\, \ldots$$

and let

$$Q = \langle t \rangle$$

be infinite cyclic. Let

$$\varphi : Q \longrightarrow \text{Aut } A$$

be defined by

$$t\varphi : a_i \longmapsto a_{i+1} \qquad (i \in \mathbf{Z})\,.$$

Form

$$E = A \rtimes Q\,.$$

Prove that E is a 2-generator group and that E' is free abelian of infinite rank.

(3)   Let R be any ring with 1 and let

$$A = R^+$$

be the additive group of R. Let Q be any subgroup of the group of units of R and let

$$\varphi : Q \longrightarrow \text{Aut } A$$

be defined by

$$q \longmapsto (q\varphi : a \longmapsto aq) \ .$$

Form

$$E = A \rtimes Q \ .$$

(i) If $R = \mathbf{Z}[x, x^{-1}]$ is the group ring of the infinite cyclic group and $Q = \text{gp}(x)$, prove that E is isomorphic to the group in (2).

(ii) Let $R = \mathbf{Z}[\frac{1}{6}]$. Then $t = \frac{2}{3}$ is a unit in R. Let $Q = \text{gp}(t)$. Is E finitely generated? Find a presentation of E.

(4) Let the group H act on a set Y and let the group Q act on a set X. Form

$$A = H^X = \{ f : X \longrightarrow H \} \ .$$

A becomes a group under coordinate-wise multiplication, and Q acts on A

$$q : f \longmapsto fq,$$

where

$$fq(x) = f(xq^{-1}) \qquad (x \in X) \ .$$

We term the semidirect product $A \rtimes Q$ a wreath product of H by Q. Notice that $A \rtimes Q$ acts on

$$X \times Y$$

by

$$(x, y)(q, f) = (xq, \, y\, f(xq)) \ .$$

If we replace H^X by $H^{(X)}$, the set of all functions from X to H which are almost always 1, we get a different group, the restricted wreath product of H by Q.

We denote the first wreath product by

$$\overline{W} = H \,\bar{\wr}\, Q$$

and the second by

$$W = H \wr Q \ .$$

(i) Let G be a group, $H \leq G$, X a right transversal of H in G. Let H act on H via the right regular representation and let γ be Frobenius' representation of G on X. Let

$$Q = G\gamma \ .$$

Verify that Cayley's representation yields via γ a faithful representation of G in

$$\overline{W} = H \,\bar{\wr}\, Q \ .$$

Hint Use the discussion above and that relating to Frobenius' representation.

3 Subgroups of free groups are free

Our objective, in this section, is to prove the following theorem of O. Schreier (1927).

Theorem 5 *Every subgroup of a free group is free.*

Before starting out on the proof of Theorem 5, we give some examples of some groups which turn out to be free. The proofs require the criterion (iv), of Theorem 1 of Chapter 1.

Examples 2 *(1) Let Ξ be a set of non-commuting variables. Consider the group of units of the ring of formal power series, with integral coefficients, in the non-commuting variables coming from Ξ. Prove that*

$$F = \mathrm{gp}(\, 1 + \xi \mid \xi \in \Xi \,)$$

is free on $\{\, 1 + \xi \mid \xi \in \Xi \,\}$. (This is a theorem of W. Magnus.)

(2) Suppose that X is a set and that $G = \mathrm{gp}(\sigma, \tau)$ is a subgroup of S_X. Furthermore suppose X_1 and X_2 are non-empty disjoint subsets of X and that

$$X_1\sigma^m \subseteq X_2 \qquad if \quad m \neq 0,$$
$$X_2\tau^n \subseteq X_1 \qquad if \quad n \neq 0.$$

Prove that G is free on $\{\sigma, \tau\}$.

Hint The trick is to verify that if

$$w = \sigma^{m_1}\tau^{n_1}\ldots\sigma^{m_k}\tau^{n_k} \qquad (m_1, n_1, \ldots, m_k, n_k \neq 0)$$

is a reduced word in σ and τ, then

$$X_1 w \subset X_1 \qquad and \qquad X_1 w \neq X_1 .$$

The first step is to examine $X_1\sigma^m \cap X_1\sigma^{m'}$. This examination leads to the conclusion that the images of X_1 under the non-zero powers σ^m of σ are disjoint subsets of X_2. So

$$X_1 \sigma^{m_1} \neq X_2,$$

which implies that

$$X_1 w \neq X_1 .$$

It is easy to sketch a proof of Theorem 5. To this end, let F be a free group, freely generated by the set X and let H be a subgroup of F. We recall that we can identify the elements of F with reduced X-words. Our objective is to find a set Y of generators of H with the property that every reduced Y-word is different from the identity in H - it will follow then, by (iv) of Theorem 1 of Chapter 1, that H is free. If we choose any set S of representatives of the right cosets of H in F, then we have already proved that

$$Y = \{\ \delta(s,x)\ =\ sx(\overline{sx})^{-1}\ \neq 1\ \mid\ s \in S,\ x \in X\ \}$$

generates H. In order to be able to prove that reduced Y-words are different from 1 in H, we need to choose S to be a *Schreier transversal* in the sense of the following

Definition 2   *A set S of representatives of the right cosets of H in F is called a (right) Schreier transversal if*

$$x_1^{\varepsilon_1} \ldots x_{n-1}^{\varepsilon_{n-1}} \in S \qquad whenever \qquad x_1^{\varepsilon_1} \ldots x_n^{\varepsilon_n} \in S.$$

The existence of right Schreier transversals is taken care of by the following lemma.

Lemma 2 *Let F be a free group on the set X and let H be a subgroup of F. Then there exists a right Schreier transversal S of H in F.*

Proof   Define the length $l(Hf)$ of the right coset Hf of H in F by

$$l(Hf)\ =\ \min\{\,l(hf)\ \mid\ h \in H\,\}.$$

We choose the elements of S in stages. First $1 \in S$. Now we proceed by induction. Suppose representatives have been chosen for all cosets of length at most n in such a way that an initial segment of a representative is again a representative. For the right cosets of length $n + 1$ we proceed as follows. Let

$$l(Hf) = n + 1\ .$$

So there exists in Hf an element $b_1 \ldots b_{n+1}$ of length $n + 1$. Now

$$l(Hb_1 \ldots b_n)\ \leq\ n$$

and so the representative $a_1 \ldots a_m\ (m \leq n)$ of $Hb_1 \ldots b_n$ has already been chosen. Consider

$$a_1 \ldots a_m b_{n+1}\ .$$

Notice that

$$Ha_1 \ldots a_m b_{n+1} = Hb_1 \ldots b_n b_{n+1}\ .$$

So
$$l(a_1 \ldots a_m b_{n+1}) = n + 1,$$

i.e., $m = n$ and in particular
$$a_1 \ldots a_n b_{n+1}$$

is a reduced X-word. We take

$$a_1 \ldots a_n b_{n+1}$$

to be the representative of Hf. It is clear that every initial segment of $a_1 \ldots$ $\ldots a_n b_{n+1}$ is again a representative, as desired.　■

In order to complete the proof of Theorem 5, we need first to introduce some notation. If $f, g \in F$ and if

$$l(fg) = l(f) + l(g),$$

we write
$$f \wedge g;$$

this means that if f and g are given as reduced X-words, then fg is itself a reduced X-word. If, on the other hand,

$$l(fg) < l(f) + l(g)$$

we sometimes write
$$f \sqcup g,$$

expressing the fact that the last letter of f cancels the first letter of g.

We are now in a position to prove that Y freely generates H.

Lemma 3 *Let F be a free group, freely generated by X, let H be a subgroup of F and let S be a right Schreier transversal of H in F. Then the following hold.*

(i) *If $\delta(s, x) \neq 1$, then*
$$\delta(s, x) = s_\triangle x_\triangle (\overline{sx})^{-1}.$$

(ii) *If $\delta(s, x) \neq 1$, then*

$$\delta(s, x) = \delta(t, y) \quad \text{if and only if} \quad s = t \quad \text{and} \quad x = y \ .$$

(iii) *If*
$$\pi = \delta(s_1, x_1)^{\varepsilon_1} \ldots \delta(s_n, x_n)^{\varepsilon_n}$$

is a reduced Y-word in the symbols $\delta(s, x)$ (which by (ii) are distinct elements if the symbols are distinct), then

$$\pi = \ldots \triangle \, x_1^{\varepsilon_1} \triangle \cdots \triangle \, x_n^{\varepsilon_n} \triangle \ldots,$$

i.e., the x and x^{-1} in the middle of $\delta(s, x)$ and $\left(\delta(s, x)\right)^{-1}$ respectively, do not cancel in the computation of the reduced X-word representing π.

We will prove each of the parts of Lemma 3 in turn. We begin with (i).

(i) Suppose $s_{\sqcup} x$. Then

$$s = t_{\triangle} x^{-1}$$

and since S is a Schreier transversal $t \in S$, i.e., $\bar{t} = t$. Now

$$sx \left(\overline{sx}\right)^{-1} = tx^{-1} x \left(\overline{tx^{-1}x}\right)^{-1} = t \, t^{-1} = 1 \, .$$

This contradicts the assumption that $\delta(s, x) \neq 1$. The case $x_{\sqcup} \left(\overline{sx}\right)^{-1}$, or equivalently, $\overline{sx}_{\sqcup} x^{-1}$ can be handled in the same way.

(ii) By (i) $s_{\triangle} x_{\triangle} \left(\overline{sx}\right)^{-1} = t_{\triangle} y_{\triangle} \left(\overline{ty}\right)^{-1}$. If $l(s) = l(t)$, $s = t$, $x = y$. If $l(s) < l(t)$, sx is an initial segment of t. So $\overline{sx} = sx$ and therefore $sx \left(\overline{sx}\right)^{-1} = 1$, a contradiction. Similarly, the remaining case cannot occur.

(iii) Note that on computing any of the reduced Y-words

$$\left(\delta(s, x)\right)^{\pm 1} \left(\delta(t, y)\right)^{\pm 1}$$

we have four possibilities

$$\ldots \triangle \, x \, \triangle \, \ldots \triangle \, y \, \triangle \, \ldots,$$
$$\ldots \triangle \, x \, \triangle \, \ldots \triangle \, y^{-1} \, \triangle \, \ldots,$$
$$\ldots \triangle \, x^{-1} \, \triangle \, \ldots \triangle \, y \, \triangle \, \ldots,$$
$$\ldots \triangle \, x^{-1} \, \triangle \, \ldots \triangle \, y^{-1} \, \triangle \, \ldots.$$

The point is that the letters exhibited here never cancel. This establishes the form of π.

This completes the proof of Theorem 5.

To sum up, suppose that the group F is free on the set X and that $H \leq F$. If we choose a right Schreier transversal S of H in F, then H is freely generated by

$$Y = \{ \; \delta(s, x) \; = \; sx(\overline{sx})^{-1} \; \neq 1 \mid s \in S, \; x \in X \; \} \, .$$

Examples 3 *Let F be free on $\{x, y\}$.*

(1) Define $\varphi : F \longrightarrow C_2 = \langle a; a^2 \rangle$ by $x \longmapsto a$, $y \longmapsto 1$. Let $H = \ker \varphi$. So $F/H \cong C_2$. Note that $F = H \cup Hx$. Then S is readily chosen: $S = \{1, x\}$.

$$Y = \{ \delta(s, \xi) \neq 1 \mid s \in S, \xi \in X \} .$$

Since $\delta(1, x) = 1$, $\delta(1, y) = y$, $\delta(x, x) = x^2$, $\delta(x, y) = xyx^{-1}$, we get

$$Y = \{ y, xyx^{-1}, x^2 \}$$

and H is free on Y.

(2) Find a set of free generators for $H = \mathrm{gp}(f^2 \mid f \in F)$. ($F/H$ is the Klein 4-group.)

(3) Define $\varphi : F \longrightarrow C_\infty = \langle a \rangle$ by $x \longmapsto a$, $y \longmapsto 1$. Let $H = \ker \varphi$. So $F/H \cong C_\infty$. Note $F = \bigcup_{i \in \mathbf{Z}} Hx^i$. Take $S = \{ x^i \mid i \in \mathbf{Z} \}$. The set Y of free generators of H we obtain is

$$Y = \{ x^i y x^{-i} \mid i \in \mathbf{Z} \} .$$

Thus a subgroup of a finitely generated free group is not always finitely generated.

(4) If G is any free group prove that G/G' is free abelian, i.e., a direct sum of infinite cyclic groups. In the case of a free group F of rank two, we find that F/F' is free abelian on xF' and yF'. So

$$S = \{ x^m y^n \mid m, n \in \mathbf{Z} \}$$

is a right Schreier transversal of F' in F. Since

$$\delta(x^m y^n, x) = x^m y^n x (\overline{x^m y^n x})^{-1} = x^m y^n x (x^{m+1} y^n)^{-1} = x^m y^n x y^{-n} x^{-(m+1)},$$
$$\delta(x^m y^n, y) = x^m y^n y (\overline{x^m y^n y})^{-1} = x^m y^{n+1} (x^m y^{n+1})^{-1} = 1,$$

F' is free on $\{ x^m y^n x y^{-n} x^{-(m+1)} \mid n \neq 0 \}$.

(5) Let φ be a homomorphism of F onto S_3. Find free generators for the kernel H of φ.

Definition 3 *Let $\mathcal{P}$ be a property of groups. Then we say a group G is virtually $\mathcal{P}$ or has $\mathcal{P}$ virtually or is virtually a $\mathcal{P}$-group, if G has a subgroup of finite index with $\mathcal{P}$.*

So a virtually finite group is finite; a virtually abelian group is a finite extension of an abelian group and, remarkably (see the remarks in the introduction), a torsion-free virtually free group is free (J.R. Stallings).

Definition 4 *Suppose H is a subgroup of a free group F. We term H a free factor of F if we can find two disjoint sets Y and Z such that Y freely generates H and $Y \cup Z$ freely generates F. If K is the subgroup generated by Z we write*

$$F = H * K$$

and term F the free product of H and K.

Notice that K is also a free factor of F.

Definition 5 *Let G be a group acting on a set S. We say that G acts transitively on S, if given any pair of elements $s, t \in S$, there exists $g \in G$ such that*

$$s\,g = t \;.$$

Then it is easy to prove the following

Lemma 4 *Suppose G acts transitively on S. Let $s_0 \in S$ be any fixed element of S and, for each $t \in S$, let $g \in G$ be chosen so that*

$$s_0\,g = t \;.$$

Then the set R of such elements of G is a complete set of representatives of the right cosets of the stabilizer J of s_0 in G.

Proof Let $c \in G$. Consider

$$s_0\,c \;.$$

There exists $r \in R$ such that

$$s_0\,c = s_0\,r \;.$$

So $cr^{-1} \in J$, i.e.,

$$J\,c = J\,r \;.$$

Hence every right coset of J in G is represented by an element of R. Moreover if $r_1, r_2 \in R$ and

$$J\,r_1 = J\,r_2,$$

then

$$s_0\,r_1 = s_0\,r_2 \;.$$

So, by the definition of R,

$$r_1 = r_2 \;. \qquad \blacksquare$$

Our next objective is to prove the following theorem of M. Hall (see his book, cited in Chapter II):

Theorem 6 *Let H be a finitely generated subgroup of a finitely generated free group F. Then F is virtually a free product of H and some other subgroup of F.*

Proof Let F be free on X, R a right Schreier transversal of H in F. Then H is free on

$$Y = \{ \, r_1 x_1 (\overline{r_1 x_1})^{-1}, \ldots, r_n x_n (\overline{r_n x_n})^{-1} \, \} \qquad (\, n < \infty, \ r_i \in R, \ x_i \in X \,) \, .$$

Let S consist of all initial segments of the elements

$$r_1, \ \overline{r_1 x_1}, \ldots, \ r_n, \ \overline{r_n x_n} \, .$$

S is a finite subset of R.

For each $x \in X$ define

$$S(x) = \{ \, s \in S \mid \overline{sx} \in S \, \} \, .$$

Notice that $S(x)$ may well be empty. Define

$$\varphi(x) : S(x) \longrightarrow S \qquad \text{by} \qquad s \longmapsto \overline{sx}.$$

Then $\varphi(x)$ is one-to-one and can therefore be continued to a permutation, again denoted by $\varphi(x)$, of S. So, allowing x to range over X, we can view this discussion as the definition of a map from X into the set of all permutations of S and hence as a homomorphism, φ say, of F into the permutation group on S. Now let $s \in S$, $x \in X$ and suppose $s \triangle x \in S$. This means that $s \in S(x)$ and

$$s \, \varphi(x) = \overline{sx} = \ sx \, .$$

Similarly if $s \in S$, $x \in X$ and $s \triangle x^{-1} \in S$ then $sx^{-1} \in S(x)$ and

$$s \, x^{-1} \, \varphi(x) = s$$

or

$$s \, \varphi(x^{-1}) = sx^{-1} \, .$$

Now suppose $t \in S$ and write t as a reduced X-product

$$t = x_1^{\varepsilon_1} \ldots x_n^{\varepsilon_n} \, .$$

Then it follows that

$$1 \, \varphi(t) = 1 \, \varphi \big(x_1^{\varepsilon_1} \ldots x_n^{\varepsilon_n} \big) = \Big(\ldots \big(1 \, \varphi (x_1^{\varepsilon_1}) \big) \ldots \Big) \varphi (x_n^{\varepsilon_n}) = x_1^{\varepsilon_1} \ldots x_n^{\varepsilon_n} = t \, .$$

Thus φ defines a transitive action of F on S and by the preceding lemma, S itself
is a complete set of representatives of the right cosets of the stabilizer J of 1 under
this action. S is closed under initial segments and therefore is a right Schreier
transversal for J in F. Denote the representative of a coset Jf in S by $\tilde{f}$. Then J
is free on

$$W = \{ \, sx\,\widetilde{sx}^{-1} \neq 1 \mid s \in S \ x \in X \, \} \ .$$

Notice that if $f \in F$, then

$$\tilde{f} = 1\,\varphi(f) \ .$$

Now consider the elements of Y, i.e. the various elements of the form $r_i x_i \, (\overline{r_i \overline{x}_i})^{-1}$.
We have

$$\widetilde{r_i x_i} = 1\,\varphi(r_i x_i) = \big(1\varphi(r_i)\big)\varphi(x_i) = r_i\,\varphi(x_i) \ .$$

But by definition

$$r_i \in S(x_i) \ .$$

So

$$r_i\,\varphi(x_i) = \overline{r_i \overline{x}_i} \ .$$

In other words

$$\widetilde{r_i x_i} = \overline{r_i \overline{x}_i} \ .$$

So $Y \subseteq W$ which completes the proof of the theorem.   ∎

Exercise 2 *A non-trivial free factor H of a free group F is a normal subgroup of
F if and only if $H=F$.*

Corollary 1 (Schreier) *If H is a finitely generated normal subgroup of a non-
trivial free group F, then either $H = 1$ or H is of finite index in F (and so F is
finitely generated).*

Definition 6 *Let $\mathcal{P}$ be an algebraic property of groups. We term a group G
residually in $\mathcal{P}$, and write $G \in \mathcal{RP}$, if for each $g \in G$, $g \neq 1$, there exists a
normal subgroup N of G such that $g \notin N$ and $G/N \in \mathcal{P}$.*

Notice that $\mathcal{P} \subseteq \mathcal{RP}$ and that $\mathcal{R}\big(\mathcal{R}(\mathcal{P})\big) = \mathcal{RP}$ (i.e., $\mathcal{R}$ is an idempotent operator).

Exercise 3 *Prove that finitely generated abelian groups are residually finite.*

Theorem 7 (F.W. Levi) *Free groups are residually finite.*

Proof It is not hard to see that it is enough to consider the case of a finitely
generated free group F. Let $f \in F$, $f \neq 1$. Then $\mathrm{gp}(f)$ is a free factor of a subgroup
J of finite index j in F. Thus we can find a free set

$$f, a_1, \ldots, a_r$$

of generators of J. Now consider

$$L = \mathrm{gp}(\, J'\,, a_1, \ldots, a_r\,, f^2\,)\ .$$

Then

$$f \notin L$$

and

$$|\,J/L\,| = 2.$$

Notice that L is of index $2j$ in F. It has only finitely many conjugates in F. Let N be their intersection. Then N is a normal subgroup of F of finite index and $f \notin N$. ∎

We recall the

Definition 7 *A group G is termed hopfian if $G/N \cong G$ implies $N=1$.*

Levi's theorem can be used to prove that finitely generated free groups are hopfian on appealing to

Theorem 8 (A.I. Malcev) *A finitely generated residually finite group G is hopfian.*

Proof Suppose that

$$G/N \,\cong\, G\ .$$

This means that the number of subgroups of a given finite index j in G/N is the same as the number of subgroups of index j in G, which is finite by Theorem 3. Let

$$J_1/N\,, \ldots,\, J_r/N$$

be the subgroups of finite index j in G/N (note that each $J_i \geq N$). This means that

$$J_1\,, \ldots,\, J_r$$

are the subgroups of index j in G. Turning this around we see that this implies that every subgroup of finite index in G contains N. So

$$N \,\leq\, \bigcap_{L \leq G,\ [G:L]<\infty} L\ .$$

But G is residually finite. Hence the intersection of the subgroups of finite index in G is trivial, i.e., $N = 1$. ∎

Exercises 4   *(1)   Prove that a finite extension of a residually finite group is residually finite.*

(2)   A cyclic extension of a finitely generated residually finite group is residually finite.

Examples 4   *Let $A = \mathbf{Z}[\frac{1}{2}]^+$, i.e. the additive group of the ring $\mathbf{Z}$ of integers with $\frac{1}{2}$ adjoined. Let $B = A/\operatorname{gp}(1)$ and let*

$$C = \{ (l, m, \overline{n}) \mid l, m \in A, \ \overline{n} = n + \operatorname{gp}(1) \in B \} .$$

Define a multiplication in C as follows:

$$(l, m, \overline{n}) \cdot (l', m', \overline{n'}) = (l + l', \ m + m', \ \overline{n + n' - ml'}) .$$

Prove the following.

(i) C is a group which is nilpotent of class two.
(ii) $\zeta C \cong C_{2^\infty} \cong B$.
(iii) $C/\zeta C \cong A \times A$.
(iv)   The mapping

$$(l, m, \overline{n}) \longmapsto (2l, \frac{m}{2}, \overline{n})$$

is an automorphism α of C of infinite order.
(v)   Let G be the semidirect product of C by an infinite cyclic group $T = \langle t \rangle$ where t acts like α on C. Prove that G is finitely generated.
(vi)   Prove that G is not a hopfian group by verifying that

$$G/N \cong G$$

where

$$N = \operatorname{gp}\left(\left(0, 0, \overline{\tfrac{1}{2}} \right) \right) .$$

Theorem 9   *(J. Nielsen 1918)   Let F be a free group of finite rank n. Suppose that F is generated by a set X of cardinality n. Then X freely generates F.*

Proof   Suppose that F is free on Y. Let φ be a map from Y onto X. Then φ defines a homomorphism, again denoted φ, of F onto F. So

$$F/\ker \varphi \cong F .$$

But F is hopfian. Hence $\ker \varphi = 1$. Thus φ is an automorphism of F which means X freely generates F.   ∎

4   The calculus of presentations

We begin this section by looking at some examples, varying our notation for presentations a little.

Examples 5 *(1)* *Let* $G = \{\pm 1\} \subseteq \mathbf{Q}^{\bullet}$, *the multiplicative group of non-zero rational numbers. Then*

$$G = \langle\, a\,;\, a^2 = 1 \,\rangle\ .$$

Here $X = \{a\}$, $\varphi : a \longmapsto -1$.

(2) $G = \mathbf{Z}^+$, *the additive group of integers. Then*

$$G = \langle a \rangle\ .$$

Here $X = \{a\}$, $R = \emptyset$ *and* $\varphi : a \longmapsto 1$ *(or* $\varphi : a \longmapsto -1$*).*

(3) $G = \mathbf{Z}[x]^+$, *the additive group of the ring of all polynomials in* x *with integer coefficients. Then*

$$G = \langle\, a_0, a_1, \ldots\ ;\ \{\,[a_i, a_j] = 1 \mid i, j \geq 0\,\} \,\rangle\ .$$

(4) $G = S_3$, *the symmetric group of degree 3. Then*

$$G = \langle\, \sigma, \tau\,;\, \sigma^3 = \tau^2 = \sigma^{\tau}\sigma = 1 \,\rangle\ .$$

(5) Every finite group has a finite presentation which is given by its multiplication table.

(6) Let G be the semidirect product of an infinite cyclic group $A = \langle a \rangle$ *by a group* $T = \langle t; t^2 = 1 \rangle$ *of order two where t acts on A by inversion. Then*

$$G = \langle\, a, t\,;\, t^2 = 1,\, a^t = a^{-1} \,\rangle\ .$$

(7) Find a presentation for the direct product of two groups.

(8) Prove that an extension of one finitely presented group by another is finitely presentable.

(9) Let

$$G = \langle\, a, t\,;\, a^t = a^2 \,\rangle\ .$$

Prove that G is the semidirect product of

$$A = \{ \ \frac{l}{2^m} \ | \ l, m \in \mathbf{Z} \ \}$$

(thought of as a subgroup of the additive group $\mathbf{Q}^+$ of rational numbers) by an infinite cyclic group $T = \langle t \rangle$, where t acts on A by multiplication by 2.

(10) Let

$$G = \langle \ X \cup \{y\} \ ; \ y = 1 \ \rangle \ .$$

If $y \notin X$, prove that G is free on X.

Of course, as we have noted before, a given group always has infinitely many different presentations. In order to illustrate one facet of this observation, suppose that f is a function from the positive integers into the positive integers. Then the following presentation

$$\langle \ x_1, x_2, \ldots \ ; \ \{x_{f(i)} = 1 \ | \ i = 1, 2, \ldots\} \ \rangle \tag{7}$$

defines a free group. We need to introduce a definition.

Definition 8 *A presentation $\langle X; R \rangle$, where $X = \{x_1, x_2, \ldots\}$ is a countable, possibly infinite set, is termed recursive if R is a recursively enumerable subset of the free group F on X. A group is termed recursively presentable if it has a recursive presentation.*

We return to the presentation above of a free group. We emphasise that a function f, defined on the natural numbers N, is said to be recursive if we can effectively compute $f(n)$ for every $n \geq 0$, i.e., the range of f is a *recursively enumerable* set. This does not mean that the complement of the range of f in N is also recursively enumerable. A subset of N is said to be recursive if it and its complement in N are recursively enumerable. There exist recursive functions whose ranges are not recursive. Now if the function f, above, is chosen so that the complement of its range is not recursively enumerable, then (7) is still a recursive presentation of a free group. One can deduce that the normal closure N of $\{x_{f(i)} \ | \ i = 0, 1, \ldots\}$ is a recursively enumerable subset of F and that the complement of N in F is not recursively enumerable. Thus the group of this presentation has an unsolvable word problem. Of course a free group of countably infinite rank has a recursive presentation with a solvable word problem. It follows that a given group can have two radically different recursive presentations.

We shall show that if we restrict our attention to finite presentations this behaviour cannot occur. More precisely, if X, R, Y, S are finite, if

$$G = \langle X; R \rangle \text{ and } G = \langle Y; S \rangle,$$

if F_1 is the free group on X, if F_2 is the free group on Y, if M is the normal closure of R in F_1 and if N is the normal closure of S in F_2, then M (suitably interpreted) is a recursive subset of F_1 if and only if N is a recursive subset of F_2. This means that having a solvable word problem is an invariant of finite presentations.

We emphasise these remarks by formulating the following definition.

Definition 9 *Let $\langle X; R \rangle$ be a presentation where $X = \{x_1, x_2, \ldots\}$ is a countable, possibly infinite, set. Then we say $\langle X; R \rangle$ has a solvable word problem if $\mathrm{gp}_F(R)$ is a recursive subset of the free group F on X. In other words we can not only effectively list the elements of $\mathrm{gp}_F(R)$ but also those of $F - \mathrm{gp}_F(R)$.*

Lemma 5 *Let $G = \langle X; R \rangle$ be a recursive presentation of G. Then the consequences of R form a recursively enumerable set.*

Proof By assumption

$$X = \{\, x_1, x_2, \ldots \,\}$$

is a countable, possibly infinite set. And R is a recursively enumerable subset of the free group F on X. So R is recursively enumerable and hence all products of conjugates of $R \cup R^{-1}$, i.e., $\mathrm{gp}_F(R)$, is a recursively enumerable subset of the free group F. ∎

This preoccupation with presentations of this kind is no idle pursuit – as I have already noted in the introduction, they play a critical role in the subgroup structure of finitely presented groups.

Exercises 5 *(1) Prove that the additive group $\mathbf{Q}^+$ of rational numbers has a recursive presentation with a solvable word problem.*

(2) Does $\mathbf{Q}^+$ have a recursive presentation with an unsolvable word problem?

(3) Prove that the restricted direct product of all the symmetric groups of finite degree has a recursive presentation with a solvable word problem.

5 The calculus of presentations (continued)

There are four simple ways of going from one presentation of a group G to another, called *Tietze transformations*, which we formulate in the form of a proposition, where we use the notation $w(\underline{x})$ in place of $w(x_1, \ldots, x_n)$.

Proposition 1 *Let G be a group. Then the following hold:*

T1 *If $G = \langle X; R \rangle$ then $G = \langle X \dot\cup Y; R \cup \{ yw(\underline{x})^{-1} \mid y \in Y \} \rangle$ where here $w(\underline{x})$ is a reduced X-word corresponding to $y \in Y$ i.e. different y's may well have different $w(\underline{x})$'s attached to them.*

T1' *If $G = \langle X \dot\cup Y; R \cup \{ yw(\underline{x})^{-1} \mid y \in Y \} \rangle$ and R does not contain any words which involve the elements y in Y, then $G = \langle X; R \rangle$.*

T2 *If $G = \langle X; R \rangle$ and S is a set of consequences of R, then $G = \langle X; R \cup S \rangle$.*

T2' *If $G = \langle X; R \cup S \rangle$ and S is a set of consequences of R, then $G = \langle X; R \rangle$.*

T1 and T1' are called *Tietze transformations of type 1*; T2 and T2' are called *Tietze transformations of type 2*.

Once one interprets these transformations, the proofs are easy; for example, consider the proof of T1.

We have

$$G = \langle X; R \rangle .$$

So X comes with its presentation map φ. We infer, although this information is not given explicitly, that the presentation map φ^+ in $\langle X \cup Y; R \cup \{ yw(\underline{x})^{-1} \mid y \in Y \} \rangle$ is φ on X, and that

$$y \varphi^+ = w(\underline{x}\varphi) .$$

Our objective is to prove that φ^+ is onto and that the kernel of φ^+ is

$$\ker \varphi^+ = \mathrm{gp}_{F^+} (R \cup \{ yw(\underline{x})^{-1} \mid y \in Y \}) , \qquad (8)$$

where now

$$F = \langle X \rangle \quad \text{and} \quad F^+ = \langle X \cup Y \rangle .$$

It is obvious that φ^+ is onto. To prove (8) we notice first, without proof, which we leave to the reader, that

$$F^+ \text{ is free on } X \cup \tilde{Y} ,$$

where

$$\tilde{Y} = \{ yw(\underline{x})^{-1} \mid y \in Y \} .$$

We define φ^+ in stages:

$$F^+ \xrightarrow{\ \chi\ } F \xrightarrow{\ \varphi\ } G$$

where

$$\chi \mid X = id , \quad \chi \mid \tilde{Y} = \text{the trivial map.}$$

Now

$$\varphi^+ = \chi \varphi .$$

So

$$\ker \varphi^+ = \chi^{-1}\big(\varphi^{-1}(1)\big) = \chi^{-1}\big(\mathrm{gp}_F(R)\big) = \mathrm{gp}_{F+}\big(R \cup \{\, yw(\underline{x})^{-1} \,|\, y \in Y \}\big) \;. \quad \blacksquare$$

Theorem 10 (Tietze) *Let*

$$G = \langle\, X; R \,\rangle \ and\ also\ G = \langle\, Y; S \,\rangle \;.$$

Then the first presentation can be transformed into the second by Tietze transformations.

Proof Let φ, ψ be the presentation maps involved in the two presentations. Then

$$y\psi = w(\underline{x}\varphi) \quad \text{for each} \quad y \in Y \;.$$

So, by $T1$,

$$G = \langle\, X \,\dot\cup\, Y \,;\, R \cup \{\, yw(\underline{x})^{-1} \,|\, y \in Y \} \,\rangle \;.$$

Now each $s \in S$ is a relator in this presentation. So, by $T2$,

$$G = \langle\, X \,\dot\cup\, Y \,;\, R \cup \{\, yw(\underline{x})^{-1} \,|\, y \in Y \} \cup S \,\rangle \;.$$

Again

$$x\varphi = v(\underline{y}\psi) \quad \text{for each} \quad x \in X \;.$$

So, by $T2$,

$$G = \langle\, X \,\dot\cup\, Y \,;\, R \cup S \cup \{\, yw(\underline{x})^{-1} \,|\, y \in Y \} \cup \{\, xv(\underline{y})^{-1} \,|\, x \in X \} \,\rangle \;.$$

We want to get rid of X. If we write $R = \{\, r(\underline{x}) \,|\, r \in R \}$, then, by $T2$,

$$G = \langle\, X \,\dot\cup\, Y \,;\, \{\, r(\underline{x}) \,|\, r \in R \} \cup S \cup \{\, yw(\underline{x})^{-1} \,|\, y \in Y \} \cup \{\, xv(\underline{y})^{-1} \,|\, x \in X \}$$
$$\cup \{\, r\big(v(\underline{y})\big) \,|\, r \in R \} \cup \{\, yw\big(v(\underline{y})\big)^{-1} \,|\, y \in Y \} \,\rangle$$

and, by $T2'$,

$$G = \langle\, X \,\dot\cup\, Y \,;\, S \cup \{\, xv(\underline{y})^{-1} \,|\, x \in X \}$$
$$\cup \{\, r\big(v(\underline{y})\big) \,|\, r \in R \} \cup \{\, yw\big(v(\underline{y})\big)^{-1} \,|\, y \in Y \} \,\rangle \;.$$

By $T1'$, we can throw away X:

$$G = \langle\, Y \,;\, S \cup \{\, r\big(v(\underline{y})\big) \,|\, r \in R \} \cup \{\, yw\big(v(\underline{y})\big)^{-1} \,|\, y \in Y \} \,\rangle \;.$$

Finally, by $T2'$,

$$G = \langle\, Y; S \,\rangle \;. \qquad\qquad \blacksquare$$

It follows immediately from the proof of this theorem that we have also proved the

Lemma 6 *Let $\langle X; R \rangle$, $\langle Y; S \rangle$ be finite presentations of the group G. Then $\langle X; R \rangle$ can be transformed into $\langle Y; S \rangle$ by a finite number of Tietze transformations, where in each instance, the transformation involved "adds" or "subtracts" a single generator and a single relator or simply adds a single relator.*

Now it is easy to prove – and the proof is left to the reader – that the following holds:

Lemma 7 *Suppose $G = \langle X; R \rangle$ is a recursive presentation of the group G and that $G = \langle X'; R' \rangle$ is a second presentation of G obtained from the first by a single Tietze transformation which adds or subtracts a single generator or a single relator. Then $\langle X'; R' \rangle$ is again a recursive presentation of G. Moreover $\langle X; R \rangle$ has a solvable word problem if and only if $\langle X'; R' \rangle$ does.*

Hence we find, on combining the previous two lemmas, that we have proved the

Theorem 11 *Two finite presentations of a group G either both have a solvable word problem or neither one does.*

Theorem 12 *(B.H. Neumann) Suppose that the group G has a finite presentation. Then every presentation of G on finitely many generators has a presentation on these generators with only finitely many of the given relators i.e. if*

$$G = \langle\, y_1, \ldots, y_m \;;\; s_1, s_2, \ldots \,\rangle \qquad (m < \infty)$$

then for, some $n < \infty$,

$$G = \langle\, y_1, \ldots, y_m \;;\; s_1, \ldots, s_n \,\rangle$$

The following simple lemma is the key to the proof of Neumann's theorem.

Lemma 8 *Suppose that*

$$G = \langle\, x_1, \ldots, x_m \;;\; r_1, \ldots, r_l \,\rangle \qquad (m < \infty, l < \infty)$$

and also

$$G = \langle\, x_1, \ldots, x_m \;;\; s_1, s_2, \ldots \,\rangle .$$

Then, for some $n < \infty$,

$$G = \langle\, x_1, \ldots, x_m \;;\; s_1, \ldots, s_n \,\rangle .$$

Proof Let φ be the underlying presentation map, F the free group on $x_1, \ldots, x_m$. Then

$$\mathrm{gp}_F(r_1, \ldots, r_l) = \ker \varphi = \mathrm{gp}_F(s_1, s_2, \ldots) .$$

So each of r_i can be expressed as a product of conjugates of the s_j and their inverses. Since only finitely many s_j come into play

$$\mathrm{gp}_F(r_1,\ldots,r_l) = \mathrm{gp}_F(s_1,\ldots,s_n)$$

for some $n < \infty$.   ∎

We are now in a position to prove Neumann's theorem.

Proof We use Tietze transformations to reduce the theorem to Lemma 8. Suppose

$$G = \langle\, x_1,\ldots,x_k \,;\, r_1,\ldots,r_l \,\rangle \qquad (k < \infty, l < \infty)\,.$$

If

$$G = \langle\, y_1,\ldots,y_m \,;\, s_1, s_2,\ldots \,\rangle$$

we add the y_i to the generators for G and then remove the x_i :

$$
\begin{aligned}
G \;=\; & \langle\, x_1,\ldots,x_k, y_1,\ldots,y_m \,;\\
& \{r_i(\underline{x}) \,|\, i = 1,\ldots,l\} \cup \{y_i w(\underline{x})^{-1} \,|\, i = 1,\ldots,m\} \,\rangle\\[4pt]
=\; & \langle\, x_1,\ldots,x_k, y_1,\ldots,y_m \,;\\
& \{r_i(\underline{x}) \,|\, i = 1,\ldots,l\} \cup \{y_i w(\underline{x})^{-1} \,|\, i = 1,\ldots,m\}\\
& \cup \{x_i v(\underline{y})^{-1} \,|\, i = 1,\ldots,k\} \,\rangle\\[4pt]
=\; & \langle\, x_1,\ldots,x_k, y_1,\ldots,y_m \,;\\
& \{r_i\left(v(\underline{y})\right) \,|\, i = 1,\ldots,l\} \cup \{y_i w\left(v(\underline{y})\right)^{-1} \,|\, i = 1,\ldots,m\}\\
& \cup \{x_i v(\underline{y})^{-1} \,|\, i = 1,\ldots,k\} \,\rangle\\[4pt]
=\; & \langle\, y_1,\ldots,y_m \,;\\
& \{r_i\left(v(\underline{y})\right) \,|\, i = 1,\ldots,l\} \cup \{y_i w\left(v(\underline{y})\right)^{-1} \,|\, i = 1,\ldots,m\} \,\rangle\,.
\end{aligned}
$$

Now apply Lemma 8.   ∎

Notation Let $N \trianglelefteq G$. Then

$$d_G(N) = \min\{\, |X| \;\big|\; \mathrm{gp}_G(X) = N \,\}\,.$$

Thus another way of putting Neumann's theorem is as follows. Suppose that G is a finitely generated group, and that F is a finitely generated free group such that

$$G \cong F/K\,.$$

Then G is finitely presented if and only if $d_F(K)$ is finite. Hence the following lemma holds:

Lemma 9 *Let G be a finitely presented group and suppose that*

$$G \cong H/N,$$

where H is finitely generated. Then

$$d_H(N) < \infty \, .$$

So we find

Corollary 1 *Let H be a finitely generated group, N a subgroup of the center of H which is not finitely generated. Then H/N is not finitely presented.*

The point here is that $d_H(N)$ is simply the minimum number $d(N)$ of generators of N.

Exercise 6 (P. Hall) *Let H be the following subgroup of $GL(3,\mathbf{R})$:*

$$H = \mathrm{gp} \left(\quad \tau = \begin{pmatrix} 1 & 0 & 0 \\ 0 & \pi & 0 \\ 0 & 0 & 1 \end{pmatrix}, \quad \alpha = \begin{pmatrix} 1 & 0 & 0 \\ 1 & 1 & 0 \\ 1 & 1 & 1 \end{pmatrix} \right) .$$

Prove
(i) ζH is free abelian of infinite rank;
(ii) $H/\zeta H \cong C_\infty \wr C_\infty = W$.
(iii) Deduce that W is not finitely presented.

Next we record another useful fact:

Theorem 13 (W. Dyck) *Suppose*

$$G = \langle \, X; R \, \rangle$$

and

$$H = \langle \, X; R \cup S \rangle \, .$$

If γ and ϑ are the respective presentation maps, then

$$x\gamma \longmapsto x\vartheta \qquad (x \in X)$$

defines a homomorphism of G onto H.

The proof is left to the reader.

Exercises 7 *(1) Let $G = \langle x, y \, ; \, x^2 = y^2 = 1 \rangle$.*
(i) Prove that G maps onto C_2.
(ii) Prove that G is infinite.

(2) Let $\langle x_1, \ldots, x_m; r_1, \ldots, r_n \rangle$ be a finite presentation of G. If $m > n$ prove that G has an infinite cyclic factor group.

(3) Let $A = \langle X; R \rangle$, $B = \langle Y; S \rangle$, where X and Y are assumed to be disjoint sets. We define the free product G of A and B, which we denote by $A * B$, to be the group given by the the following presentation:

$$G = \langle X \cup Y; R \cup S \rangle \, .$$

Prove
(i) A and B embed in G.
(ii) If $a \in A$, $a \neq 1$, and $b \in B$, $b \neq 1$, then $[a, b] \neq 1$.
    (Hint: Map G onto $A \wr B$.)
(iii) If A and B are free, so is $A * B$.
(iv) If $A * B$ is free, so are A and B.
(v) For every group H and every pair of homomorphisms α of A into H and β of B into H, there exists a unique homomorphism of G into H which agrees with α on A and β on B.

6 The Reidemeister-Schreier method

Suppose that

$$G = \langle X; R \rangle \tag{9}$$

and that

$$H \leq G \, . \tag{10}$$

Now (9) comes with an underlying presentation map γ, say. So if F is the free group on X, then γ induces an isomorphism

$$\gamma_* \; : \; F/K \overset{\sim}{\longrightarrow} G \tag{11}$$

where

$$K = \mathrm{gp}_F(R) \, . \tag{12}$$

Notice then that

$$\gamma_*^{-1}(H) = E/K \tag{13}$$

where E is a subgroup of F. So (13) is a presentation of H in the making. Indeed let T be a Schreier transversal of E in F. Then E is free on Y, where

$$Y = \{ \, \delta(t, x) = tx(\overline{tx})^{-1} \neq 1 \mid t \in T, \; x \in X \, \} \, . \tag{14}$$

Now (13) tells us that

$$\gamma_* \,|\, E \;:\; E/K \xrightarrow{\sim} H \,. \tag{15}$$

So, by (14),

$$\gamma \,|\, Y \quad \text{is a presentation map of } H. \tag{16}$$

Indeed, by (15),

$$H = \langle Y; S \rangle$$

where we have to properly interpret S.

Now, by (12),

$$K = \mathrm{gp}_F(R) = \mathrm{gp}_E\left(\{\, trt^{-1} \mid t \in T,\, r \in R \,\}\right) \,.$$

But the elements trt^{-1} come to us as X-words, not Y-words. However

$$K \leq E,$$

and so each trt^{-1} can be expressed as a reduced Y-word. Let us denote this expression for trt^{-1} by $\varrho(trt^{-1})$. Then what we mean by S above is precisely this set of all rewritten X-words:

$$S = \{\, \varrho(trt^{-1}) \mid t \in T,\, r \in R \,\} \,.$$

To repeat then, we have a presentation of H:

$$H = \langle\, \{\, \delta(t, x) \neq 1 \mid t \in T,\, x \in X \,\} \,;\, \{\, \varrho(trt^{-1}) \mid t \in T,\, r \in R \,\} \,\rangle \,. \tag{17}$$

Examples 6 Let $G = \langle b, u\,;\, ubu^{-1} = b^2 \rangle$.
By Dyck's theorem, G maps onto C_∞:

$$C_\infty = \langle b, u\,;\, ubu^{-1} = b^2,\, b = 1 \rangle \,.$$

Let H be the kernel of this map. Then G/H is infinite cyclic with generator uH. So $G/H = \bigcup_{n \in \mathbf{Z}} u^n H$. Thus, if we go back to the ambient free groups E and F in the discussion of the Reidemeister-Schreier method above, we find that:

(i) $\;\; T = \{\, u^n \mid n \in \mathbf{Z} \,\}$;
(ii) $\;\; Y = \{\, u^n b u^{-n}\, (= b_n) \mid n \in \mathbf{Z} \,\}$;
(iii) $S = \{\, \varrho(u^n (ubu^{-1} b^{-2}) u^{-n}) \mid n \in \mathbf{Z} \,\} = \{\, b_{n+1} b_n^{-2} \mid n \in \mathbf{Z} \,\}$.

Thus

$$H = \langle\, \dots, b_{-1}, b_0, b_1, \dots \,;\, \dots, b_0 = b_{-1}{}^2,\, b_1 = b_0{}^2,\, b_2 = b_1{}^2, \dots \,\rangle \,.$$

It follows by means of Tietze transformations of type 1 that

$$H = \langle \ldots, b_{-1}, b_0 \; ; \; \ldots \; b_{-1} = b_{-2}{}^2, b_0 = b_{-1}{}^2 \rangle \; .$$

Now let's map H into $\mathbf{Q}^+$ as follows:

$$b_n \longmapsto 2^n, \qquad n = 0, -1, -2, \ldots \; .$$

By Dyck's theorem this defines a homomorphism μ of H into $\mathbf{Q}^+$. Indeed the image of H is simply

$$D = \left\{ \frac{l}{2^n} \; \middle| \; l, n \in \mathbf{Z} \right\} \; .$$

It is easy to see that μ is an isomorphism because D is a union of infinite cyclic groups. Notice that

$$D = \mathbf{Z} \left[\frac{1}{2} \right]^+ \; .$$

Exercises 8 *(1) Let*

$$G = \langle \, a, b, c, d \, ; \, aba^{-1}b^{-1}cdc^{-1}d^{-1} = 1 \, \rangle \; .$$

Find, for each positive integer n, a presentation for

$$H_n = \mathrm{gp}_G(a^n, b, c, d).$$

(Hint: H_n has a presentation of the form

$$H_n = \langle \, a_1, b_1, \ldots, a_g, b_g \; ; \; \prod_{i=1}^{g} a_i b_i a_i^{-1} b_i^{-1} = 1 \, \rangle \; .$$

Find g.)

(2) Let G be as in (1) and let

$$H = \mathrm{gp}_G(b, c, d) \; .$$

Prove that H is free.

(3) Let

$$G = \langle \, a, b \, ; \, a^3 = b^3 = (ab)^3 = 1 \, \rangle \; .$$

Prove that G' is free abelian of rank two, and hence that G is infinite.

7   Generalized free products

Suppose that

$$\{ G_i = \langle X_i \, ; \, R_i \rangle \, | \, i \in I \}$$

is an indexed family of presentations $G_i = \langle X_i \, ; \, R_i \rangle$ of the groups G_i, and suppose that H is another group equipped with monomorphisms

$$\varphi_i \; : \; H \longrightarrow G_i \qquad (i \in I) \, .$$

Then we term the group G defined by the presentation

$$G = \langle \, \bigcup_{i \in I} X_i \, ; \, \bigcup_{i \in I} R_i \bigcup \{ h \varphi_i h^{-1} \varphi_j \mid h \in H, i, j \in I \} \, \rangle \qquad (18),$$

where we assume here that the X_i are disjoint, the *generalized free product of the G_i amalgamating H*. We express the fact that G is given by such a presentation by writing

$$G = \underset{\substack{H \\ i \in I}}{\Large *} \, G_i \quad \text{or} \quad \text{sometimes} \quad G = \prod_{i \in I}^{*} \{ G_i ; H \}.$$

If $H = 1$, then G is termed the *free product of the G_i* and we write

$$G = \underset{i \in I}{\Large *} \, G_i \quad \text{or} \quad \text{sometimes} \quad G = \prod_{i \in I}^{*} G_i.$$

If $I = \{1, \ldots, n\}$ is finite, we sometimes denote G by

$$G = G_1 \underset{H}{*} G_2 \underset{H}{*} \ldots \underset{H}{*} G_n \quad \text{or} \quad G = \{ \prod_{i=1}^{n}{}^{*} G_i ; H \},$$

or if $H = 1$ by

$$G = G_1 * G_2 * \ldots * G_n \quad \text{or} \quad G = \prod_{i=1}^{n}{}^{*} G_i.$$

When $n = 2$, this reduces to

$$G = G_1 \underset{H}{*} G_2 \quad \text{or} \quad G = \{ G_1 * G_2 ; H \},$$

or again, when $H = 1$, by

$$G = G_1 * G_2 \, .$$

We assume that in (18) each $h\varphi_i$ is expressed as an X_i-product so that (18) actually looks like a presentation. According to Dyck's theorem there are canonical homomorphisms μ_i of each G_i to G. It turns out that the μ_i are monomorphisms and that if we identify G_i with $G_i\mu_i$, then $h\varphi_i = h\varphi_j$ for all $h \in H$, $i, j \in I$. So we can identify H with any one of its images $H\varphi_i$ (which we have already identified with $H\varphi_i\mu_i$) and it then turns out that

$$G_i \cap G_j = H \qquad (i, j \in I,\ i \neq j)\,,$$

i.e., the G_i intersect in precisely H. Notice that

$$G = \mathrm{gp}\Big(\bigcup_{i \in I} G_i \Big)\,.$$

It follows that every element $g \in G$ can be expressed in the form

$$g = y_1 \ldots y_n h \qquad (n \geq 0) \tag{19}$$

where

(i) $y_j \in G_{i_j} - H$ and $i_j \neq i_{j+1}$ for $j = 1, \ldots, n-1$;

(ii) $h \in H$.

In particular if $g \notin H$, then (19) can be rewritten as

$$g = z_1 \ldots z_n \qquad (n \geq 1) \tag{20}$$

where

(iii) $z_j \in G_{i_j} - H$ and $i_j \neq i_{j+1}$ for $j = 1, \ldots, n-1$.

Theorem 14 (O. Schreier) *Suppose that $G = \{\prod^* G_i; H\}$. Then*
(a) the G_i embed into G;
(b) $G_i \cap G_j = H$ $(i, j \in I,\ i \neq j)$;
(c) every product of the form (20) satisfying (iii) is not equal to 1 in G.
Conversely, suppose that G is a group and that H is a subgroup of G. Furthermore, suppose that $\{G_i \mid i \in I\}$ is an indexed family of subgroups of G, that G is generated by these subgroups G_i and that $G_i \cap G_j = H$ $(i, j \in I,\ i \neq j)$. Then

$$G = \{\prod_{i \in I}{}^* G_i; H\}$$

if and only if every product of the form (20) satisfying (iii) is not equal to 1 in G. (For a proof see, e.g., the book A.G. Kurosh, Vol. 2, cited in Chapter 1).

Corollary 1   *Let*

$$G = \{\prod_{i \in I}{}^* G_i; H\}$$

and suppose that $F_i \leq G_i$ $(i \in I)$, *that* $K \leq H$ *and that*

$$F_i \cap F_j = K \qquad (i, j \in I, \ i \neq j) \ .$$

If F is the subgroup of G generated by the F_i, then

$$F = \{ \prod_{i \in I}^* F_i; K \}$$

In particular, if $K = 1$, then

$$F = \prod_{i \in I}^* F_i \ .$$

Furthermore, if each F_i is infinite cyclic, then F is free.

This corollary helps to explain the existence of many free subgroups in generalized free products.

The importance of generalized free products lies in the fact that they occur frequently in a variety of different contexts, and, furthermore, they can be used to construct important classes of groups.

Chapter IV
Recursively presentable groups, word problems and some applications of the Reidemeister-Schreier method

1 Recursively presentable groups

The following lemma is due to G. Higman.

Lemma 1 *A finitely generated subgroup of a finitely generated recursively presentable group is recursively presentable.*

Proof Let $G = \langle X; R \rangle$ be a recursive presentation of the group G with X finite. So

$$G \cong F/K$$

where F is free on X and $K = \mathrm{gp}_F(R)$. As we have already noted before, K is a recursively enumerable subset of F. Let H be a finitely generated subgroup of G. Then

$$H \cong EK/K$$

where E is a finitely generated subgroup of F. So

$$H \cong E/(E \cap K).$$

Now E is a recursively enumerable subset of F and hence so too is $E \cap K$. If we now enumerate the subset $E \cap K$ of F and simultaneously enumerate the elements of E, we can recursively enumerate the elements of K, each written as a product of the generators of E. This means that $E \cap K$ is a recursively enumerable subset of E. Therefore H is recursively presentable. ∎

Corollary 1 *A finitely generated subgroup of a finitely presented group is recursively presentable.*

Lemma 2 *Let G be a finitely generated subgroup of $\mathrm{GL}(n,K)$, where K is any commutative field and n is any positive integer. Then G has a recursive presentation with a solvable word problem.*

Proof   Suppose that G is generated by the finitely many matrices

$$M_1, \ldots, M_q.$$

It follows that G can be viewed as a group of matrices over a finitely generated subfield L of K. Now L can be embedded in the algebraic closure U of

$$P(x_1, x_2, \ldots),$$

the field of fractions of the polynomial ring in infinitely many variables $x_1, x_2, \ldots$ over the prime field P of K. It follows from a theorem of M.O. Rabin that U is a *computable field*, i.e., we can effectively list the elements of U and effectively compute sums and products and scalar multiples of elements in U.

We now view the generators $M_1, \ldots, M_q$ of G as being matrices with coefficients in U. So if we recursively list all of the words in $M_1, \ldots, M_q$, we can simultaneously determine their values in $GL(n,K)$ and hence recursively enumerate those words which reduce to the identity. It follows that G has a recursive presentation with a solvable word problem.   ∎

In fact we have proved the

Lemma 3   *Let U be a countable computable field. Then $GL(n, U)$ is recursively presentable.*

Of course $GL(n, U)$ is nothing but the group of automorphisms of a finite dimensional vector space over U. An analogous result, due to Baumslag, Cannonito and Miller, holds also for automorphism groups of finitely presented groups.

Theorem 1   *Let G be a finitely presented group. Then $\operatorname{Aut} G$ is recursively presentable. If G has a solvable word problem then $\operatorname{Aut} G$ has a presentation with a solvable word problem.*

Proof   Suppose

$$G = \langle x_1, \ldots, x_m \, ; \, r_1(x_1, \ldots, x_m) = 1, \ldots, r_n(x_1, \ldots, x_m) = 1 \rangle \qquad (m, n \le \infty).$$

We first list all automorphisms of G. To do so, notice that if $\alpha \in \operatorname{Aut} G$,

$$x_i \alpha = w_i(x_1, \ldots, x_m) \qquad \text{for suitable words } w_i \qquad (i = 1, \ldots, m). \qquad (1)$$

Now α is a homomorphism. So

$$r_j\big(\, w_1(x_1, \ldots, x_m), \ldots, w_m(x_1, \ldots, x_m) \,\big) = 1 \qquad (j = 1, \ldots, n). \qquad (2)$$

Suppose $\beta : G \longrightarrow G$ is the inverse of α. Then

$$x_i\beta = v_i(x_1,\ldots,x_m) \qquad \text{for suitable words } v_i \qquad (i = 1,\ldots,m) . \qquad (3)$$

Again

$$r_j\big(v_1(x_1,\ldots,x_m),\ldots,v_m(x_1,\ldots,x_m)\big) = 1 \qquad (j = 1,\ldots,n) . \qquad (4)$$

Moreover, since α and β are inverses,

$$\begin{aligned} v_i\big(w_1(x_1,\ldots,x_m),\ldots,w_m(x_1,\ldots,x_m)\big) &= x_i \qquad (i = 1,\ldots,m), \\ w_i\big(v_1(x_1,\ldots,x_m),\ldots,v_m(x_1,\ldots,x_m)\big) &= x_i \qquad (i = 1,\ldots,m). \end{aligned} \qquad (5)$$

Conversely if $(w_1,\ldots,w_m)$ and $(v_1,\ldots,v_m)$ are pairs of m-tuples of words in $x_1,\ldots,x_m$ satisfying (2), (4) and (5), then the maps given in (1) and (3) are automorphisms of G which are inverses. Since G is finitely presented, it follows that we can recursively enumerate Aut G. Notice also that if, in addition, G has a solvable word problem, then Aut G is a recursive subset of the set of all mappings of G into G. We now take the entire set Aut G of automorphisms of G as a set of generators of Aut G. Now since G is finitely presented, the set of such pairs is recursively enumerable, and if G has a solvable word problem this set of pairs is even recursive. Words in this set of generators of Aut G act on the generators $x_1,\ldots,x_m$ of G, and a given word takes the value 1 if and only if it leaves every x_i fixed. In more detail, let w be such a word. Then

$$x_i w = u_i(x_1,\ldots,x_m) \qquad (i = 1,\ldots,m) .$$

So $w = 1$ in Aut G if and only if

$$u_i(x_1,\ldots,x_m) =_G x_i \qquad (i = 1,\ldots,m) . \qquad (6)$$

But the set of such equations is recursively enumerable. So Aut G is recursively presentable. If in addition G has a solvable word problem we can actually decide whether or not (6) holds. In this case it then follows that Aut G has a solvable word problem. ∎

2 Some word problems

Here we consider two classes of groups which have solvable word problem. The first of them is due to McKinsey:

Theorem 2 *A finitely presented residually finite group has a solvable word problem.*

Proof Suppose that G is a finitely presented, residually finite group given by the finite presentation

$$G = \langle\, x_1,\ldots,x_m\,;\, r_1(x_1,\ldots,x_m) = 1,\ldots,r_n(x_1,\ldots,x_m) = 1 \,\rangle .$$

We can enumerate all of the homomorphisms of G into all finite groups. Indeed for each symmetric group S_l , $l = 1, 2, \ldots$, we first enumerate the finitely many m-tuples $(\overline{x}_1, \ldots, \overline{x}_m) \in (S_l)^m$ of elements of S_l and determine which of them satisfy the equations

$$r_j(\overline{x}_1, \ldots, \overline{x}_m) = 1 \qquad (j = 1, \ldots, n) .$$

For those which do, the mappings $x_i \longmapsto \overline{x}_i$ $(i = 1, \ldots, m)$ define homomorphisms

$$G \longrightarrow \mathrm{gp}(\overline{x}_1, \ldots, \overline{x}_m) \leq S_l , \qquad x_i \longmapsto \overline{x}_i .$$

At the same time we can enumerate all consequences of the given defining relators of G. It follows from the residual finiteness of G that if w is any $\{x_1, \ldots, x_m\}$-word, then either we will find $w =_G 1$ or else that $w\varphi \neq 1$ for some homomorphism φ of G into some S_l. This solves the word problem for G. ∎

The next theorem is due to Kuznetsov:

Theorem 3 *A finitely generated, recursively presented, simple group G has a solvable word problem. (Note that by definition $G \neq 1$.)*

Proof Suppose that G is given by the recursive presentation

$$G = \langle\, x_1, \ldots, x_m \,;\, r_1, r_2, \ldots \,\rangle \qquad (m < \infty) .$$

To determine whether a reduced $\{x_1, \ldots, x_m\}$-word w has the value 1 in G, enumerate the consequences of $r_1, r_2, \ldots$. At the same time enumerate the consequences of $w, r_1, r_2, \ldots$. If w appears in the first list, $w = 1$ in G. If all of $x_1, \ldots, x_m$ appear in the second list, then $w \neq 1$ in G. This algorithm solves the word problem for G. ∎

3 Groups with free subgroups

In 1972 J.Tits (*Free subgroups in linear groups*, **J. Algebra 20** (1972), 250-270) proved the following

Theorem 4 *A finitely generated group of matrices over a commutative field is either virtually solvable or contains a free subgroup of rank two.*

This has led to what has now come to be known as the *Tits alternative*, in accordance with the following

Definition 1 *A group satisfies the Tits alternative if either it is virtually solvable, or contains a free subgroup of rank two.*

Unlike matrix groups, finitely generated groups need not satisfy the Tits alternative – the Grigorchuk-Gupta-Sidki group, for instance, is such an example. Finitely presented examples are harder to find, but they do exist: Groups of piecewise linear homeomorphisms of the real line, Matthew G. Brin & Craig C. Squier, **Invent. math. 79** (1985), 485-498 and C.H. Houghton (unpublished). Here is one of the Brin, Squier examples: Let G be generated by the permutations τ and σ of the real line, defined as follows:

$$\tau \; : \; x \longmapsto x + 1 \, ,$$

$$\sigma \; : \; x \longmapsto \begin{cases} x & (-\infty < x \le 0) \\ 2x & (0 \le x \le 1) \\ x + 1 & (1 \le x < \infty). \end{cases}$$

Then G turns out to be a finitely presented group which does not satisfy the Tits alternative.

The question therefore arises as to which finitely presented groups satisfy the Tits alternative. I want to dicuss next some answers to this question which fall out of the Reidemeister-Schreier method.

First let me recall that if

$$G = \langle X; R \rangle \, , \; H = \langle Y; S \rangle \, ,$$

(where X and Y are disjoint sets) and if C is a group equipped with two monomorphisms

$$\vartheta \; : \; C \longrightarrow G \quad , \quad \varphi \; : \; C \longrightarrow H \, ,$$

then we term

$$K = \langle X \cup Y; R \cup S \cup \{ c\vartheta c^{-1}\varphi \, | \, c \in C \} \rangle$$

the generalized free product of G and H amalgamating C and denote it by

$$K = G \underset{C}{*} H \, .$$

The canonical homomorphisms of G and H into K are monomorphisms, and if we identify G and H with their images in K then

$$G \cap H = C \, .$$

The point here is the following

Lemma 4 *If $C \neq G$ and $C \neq H$ and C has index at least three in one or other of G and H, then $G \underset{C}{*} H$ contains a free subgroup of rank two.*

The proof of this lemma is left as an exercise.

Next I want to introduce the notion of an HNN extension, taking for granted, for the moment, some its properties (see Chapter VI for more details).

Definition 2 *Let*

$$B = \langle X; R \rangle$$

be a group given by a presentation and suppose U and V are subgroups of B equipped with an isomorphism

$$\tau \; : \; U \xrightarrow{\sim} V \, .$$

Then we term

$$E = \langle X, t; \, R \cup \{ tut^{-1} = u\tau \, | \, u \in U \} \rangle \qquad (\text{where } t \notin X)$$

an HNN extension of B with stable letter t, associated subgroups U and V and associating isomorphism τ.

It turns out that
(i)   B embeds in E.
(ii) t is of infinite order in E and U and V are conjugate in E via t.
We refer to B as the *base group* of E and sometimes write

$$E = \langle B, t; \, tUt^{-1} = V \rangle \, .$$

E will be termed an *ascending HNN extension* if either U=B or V=B.

The following definition is due to G. Higman.

Definition 3 *A group G is termed indicable if there exists a homomorphism of G onto the infinite cyclic group.*

Notice that a finitely generated group G is indicable if and only if G_{ab} is infinite.

We need one additional piece of notation. To this end, suppose that X is a subset of some group, and that

$$g = x_1^{\varepsilon_1} \ldots x_n^{\varepsilon_n} \qquad (x_i \in X, \, \varepsilon_i = \pm 1)$$

is an X-word. Then we define, for each $x \in X$,

$$\exp_x g = \sum_{x_i = x} \varepsilon_i$$

and term $\exp_x g$ the *exponent sum of x* in the X-word g.

The following theorem, which was proved by Bieri and Strebel in 1978, may be viewed as a version of the Tits alternative.

Theorem 5 *Suppose that G is a finitely presented indicable group. Then either G is an ascending HNN extension with a finitely generated base group or else G contains a free subgroup of rank two.*

Proof Let μ be a homomorphism of G onto the infinite cyclic group. So if $H = \ker \mu$, G/H is infinite cyclic on, say, Ht $(t \in G)$. Now G is finitely generated. So we can find a set of generators for G of the form

$$t, a_1, \ldots, a_m \qquad (a_i \in H) \qquad\qquad (m < \infty) \ .$$

It follows from Neumann's theorem that we can present G finitely using these generators:

$$G = \langle t, a_1, \ldots, a_m \, ; \, r_1, \ldots, r_n \rangle \qquad (n < \infty) \ .$$

It follows that

$$H = \mathrm{gp}_G(a_1, \ldots, a_m),$$

and that

$$\exp_t r_k = 0 \qquad (k = 1, \ldots, n) \ .$$

Thus $G = \bigcup_{i \in \mathbf{Z}} Ht^i$. Hence $\{t^i \mid i \in \mathbf{Z}\}$ is a right Schreier transversal for H in G. It is easy then to apply the Reidemeister-Schreier process to obtain a presentation for H. To this end put

$$t^i a_j t^{-i} = a_{j,i} \qquad (j = 1, \ldots, m, \ i \in \mathbf{Z})$$

and put the rewrite

$$\varrho(t^i r_k t^{-i}) = r_{k,i} \qquad (k = 1, \ldots, n, \ i \in \mathbf{Z}) \ .$$

We need to look more closely at the rewrites $\varrho(r_k)$ of the r_k. Since we lose nothing by replacing each r_k by a conjugate of itself by a power of t, we may assume that for $k = 1, \ldots, n$,

$$\varrho(r_k) = r_{k,0} = r_k(a_{1,0}, \ldots, a_{1,\lambda}, \ldots, a_{m,0}, \ldots, a_{m,\lambda}) \ .$$

Here we use λ to denote a positive integer, chosen once and for all. Our notation is functional and is designed to reflect the fact that each r_k is some product of the generators listed without carrying with it the further implication that all of the generators listed actually appear. It follows that

$$r_{k,i} = r_k(a_{1,i}, \ldots, a_{1,\lambda+i}, \ldots, a_{m,i}, \ldots, a_{m,\lambda+i})$$

where $k = 1, \ldots, n, \ i \in \mathbf{Z}$. Hence if we put

$$Y = \{ a_{j,i} \mid j = 1, \ldots, m, \ i \in \mathbf{Z} \},$$

and
$$R = \{\, r_{k,i} \mid k = 1, \ldots, n, \ i \in \mathbf{Z} \,\},$$
then
$$H = \langle Y; R \rangle .$$
Put
$$H^+ = \mathrm{gp}(\, a_{j,i} \mid j = 1, \ldots, m, \ i \geq 0)$$
and
$$H^- = \mathrm{gp}(\, a_{j,i} \mid j = 1, \ldots, m, \ i \leq \lambda) .$$
Finally we put
$$U = \mathrm{gp}(\, a_{j,i} \mid j = 1, \ldots, m, \ 0 \leq i \leq \lambda) .$$

So U is a finitely generated subgroup of both H^+ and H^-. We now form the generalized free product $\widetilde{H}$ of H^+ and H^- amalgamating U:

$$\widetilde{H} = H^+ \underset{U}{*} H^- .$$

In order to avoid confusion we present $\widetilde{H}$ using different letters to those previously used. First we present H^+:

$$H^+ = \langle \, \{\, x_{j,i} \mid j = 1, \ldots, m, \ i \geq 0 \} \, ;$$
$$\{\, r_k(x_{1,i}, \ldots, x_{1,\lambda+i}, \ldots, x_{m,i}, \ldots, x_{m,\lambda+i}) \mid k = 1, \ldots, n, \ i \geq 0 \} \cup S^+ \, \rangle$$

where S^+ is a set of additional relations, if needed, to define H^+ in the manner indicated. Next we present H^-:

$$H^- = \langle \, \{\, y_{j,i} \mid j = 1, \ldots, m, \ i \leq \lambda \} \, ;$$
$$\{\, r_k(y_{1,i}, \ldots, y_{1,\lambda+i}, \ldots, y_{m,i}, \ldots, y_{m,\lambda+i}) \mid k = 1, \ldots, n, \ i \leq 0 \} \cup S^- \, \rangle$$

where S^- is defined similarly to S^+. So, adopting the obvious notation,

$$H^+ = \langle X; R^+ \cup S^+ \rangle , \qquad H^- = \langle Y; R^- \cup S^- \rangle .$$

Hence

$$\widetilde{H} = \langle X \cup Y; R^+ \cup S^+ \cup R^- \cup S^- \cup \{ x_{j,i} = y_{j,i} \mid j = 1, \ldots, m, \ 0 \leq i \leq \lambda \} \rangle .$$

The point of all of this is to help justify the claim that

$$\widetilde{H} \cong H . \tag{7}$$

First we concoct a homomorphism $\widetilde{\sigma}$ of $\widetilde{H}$ onto H. The thing to notice is that all of the relations we have used to define $\widetilde{H}$ go over into relations in H. More precisely, suppose that we define

$$x_{j,i}\widetilde{\sigma} = a_{j,i} \ (j = 1, \ldots, m, \ i \geq 0), \ y_{j,i}\widetilde{\sigma} = a_{j,i} \ (j = 1, \ldots, m, \ i \leq \lambda).$$

Then $\tilde{\sigma}$ maps the kernel of the presentation map of $\tilde{H}$ onto the identity of H. So $\tilde{\sigma}$ induces a homomorphism of $\tilde{H}$ onto H – this is an easily verified variation of W. Dyck's theorem using different sets of generators. Similarly, the map σ,

$$a_{j,i}\sigma = x_{j,i} \ (j = 1,\ldots,m, \ i \geq 0), \ a_{j,i}\sigma = y_{j,i} \ (j = 1,\ldots,m, \ i < 0),$$

defines a homomorphism of H into $\tilde{H}$. These two homomorphisms are inverses of each other. So we have proved (7) and essentially the theorem itself. To see why, notice that we now know that (cf. (7))

$$H = H^- \underset{U}{*} H^+ .$$

There are a number of possibilities. First suppose that $U \neq H^+$, $U \neq H^-$ and that U is of index at least three in one of H^+, H^-. Then H (and therefore G) contains a free subgroup of rank two. If U is of index two in both H^+ and H^- then they are both finitely generated. Hence H is also finitely generated and G is an ascending HNN extension where H is not only the base group but the pair of associated subgroups as well. Finally, if $U = H^+$ or $U = H^-$, then it is not hard to see that G is an ascending HNN extension with base group U. We note also, for later use, that if either H^+ or H^- is finitely generated, then G is an HNN extension with a finitely generated base. ∎

The technique involved in this proof can be exploited to yield the next theorem, which is due to G. Baumslag and P.B. Shalen.

Theorem 6 *Let G be a finitely presented indicable group which is not an ascending HNN extension with a finitely generated base. Then G is virtually a non-trivial generalized free product of two finitely generated groups where the amalgamated subgroup is of infinite index in one factor and of arbitrarily large index in the other.*

Thus most finitely presented groups are virtually generalized free products of two finitely generated groups.

Proof The proof partly mimics and makes use of the proof of the Bieri-Strebel theorem just given. Thus we assume the notation used there. In particular,

$$G = \langle t, a_1,\ldots,a_m \, ; \, r_1,\ldots,r_n \rangle, \ H = \mathrm{gp}_G(a_1,\ldots,a_m),$$

where as before,

$$\exp_t r_k = 0 \qquad (k = 1,\ldots,n),$$

and G/H is infinite cyclic on Ht. Again we assume that

$$\varrho(r_k) = r_k(a_{1,0},\ldots,a_{1,\lambda},\ldots,a_{m,0},\ldots,a_{m,\lambda})$$

for some fixed integer $\lambda > 0$. Now we put

$$H_l = \text{gp}_G(t^l, a_1, \ldots, a_m)$$

where we assume only that l is a very large positive integer. Now, by assumption, G is not an ascending HNN extension with a finitely generated base. Hence if

$$C = \text{gp}(a_{j,0}, \ldots, a_{j,\lambda-1}\ a_{j,l-\lambda}, \ldots, a_{j,l-1}\ (j = 1, \ldots, m))$$

then it follows that C is of infinite index in H^+ since H^+ is not finitely generated. Thus if we put

$$H_l{}^+ = \text{gp}(a_{j,0}, \ldots, a_{j,l-1}\ (j = 1, \ldots, m)),$$

and make sure that l is large enough, then the index of C in $H_l{}^+$ can be made as large as we please.

Our next objective is to compute a presentation for $H_l{}^+$. The set $1, t, \ldots, t^{l-1}$ is a right Schreier transversal for H_l in G. It follows from the method of Reidemeister and Schreier, that we can take

$$\{t^l\} \cup \{a_{j,i} \mid j = 1, \ldots, m,\ 0 \le i \le l-1\}$$

to be the generators of H_l. The relators of H_l that come out of this process have to be examined a little more closely. They take the form

$$
\begin{aligned}
r_{k,0} \quad &= r_k(a_{1,0}, \ldots, a_{1,\lambda}, \ldots, a_{m,0}, \ldots, a_{m,\lambda}) \\
&\ \vdots \\
r_{k,l-\lambda-1} &= r_k(a_{1,l-\lambda-1}, \ldots, a_{1,l-1}, \ldots, a_{m,l-\lambda-1}, \ldots, a_{m,l-1}) \\
r_{k,l-\lambda} &= r_k(a_{1,l-\lambda}, \ldots, a_{1,l-1}, t^l a_{1,0} t^{-l}, \ldots, a_{m,l-\lambda}, \ldots, a_{m,l-1}, t^l a_{m,0} t^{-l}) \\
&\ \vdots \\
r_{k,l-1} &= r_k(a_{1,l-1}, t^l a_{1,0} t^{-l}, \ldots, t^l a_{1,\lambda-1} t^{-l}, \ldots, \\
&\qquad\quad a_{m,l-1}, t^l a_{m,0} t^{-l}, \ldots, t^l a_{m,\lambda-1} t^{-l}).
\end{aligned}
$$

Put

$$H_l{}^- = \text{gp}(t^l;\ a_{j,0}, \ldots, a_{j,\lambda-1};\ a_{j,l-\lambda}, \ldots, a_{j,l-1}\ (j = 1, \ldots, m)).$$

So $H_l{}^-$ is a $(1 + 2m\lambda)$-generator group. Notice that it follows, as in the proof of the Bieri-Strebel theorem, that

$$H_l = H_l{}^+ \underset{C}{*} H_l{}^-,$$

where C is the $2m\lambda$-generator group given above. Now t is of infinite order modulo H. Hence C is of infinite index in $H_l{}^-$. This proves the theorem. ∎

We can use the same kind of argument, together with an idea of B. Baumslag & S.J. Pride, to prove the next

Theorem 7 *Let G be a finitely presented group and suppose that*

$$W = C_h \wr C_\infty \qquad (h \geq 2) ,$$

the wreath product of a cyclic group of order h (possibly infinite) and an infinite cyclic group, is a quotient of G. Then G contains a subgroup H of finite index which maps onto the free product

$$K = C_\infty * C_h .$$

Notice that if we consider the normal closure J in K of the infinite cyclic factor, then by the Reidemeister-Schreier method, J turns out to be a free group of rank h. Since J is of index $h \geq 2$ in K it follows that we have proved the

Corollary 1 *If G satifies the hypothesis of Theorem 7, then it contains a subgroup of finite index which maps onto the free group of rank two.*

I shall say more about this in a little while, after the proof of Theorem 7.

Proof To begin with, notice that W can be generated by two elements τ and α, which have the following properties:
(i) α is of order h and τ is of infinite order;
(ii) if we put

$$\alpha_i = \tau^i \alpha \tau^{-i} \qquad (i \in \mathbf{Z}) ,$$

then $A = \mathrm{gp}_W(\alpha)$ is the direct product of the cyclic groups $A_i = \mathrm{gp}(\alpha_i)$ of order h:

$$A = \prod_{i \in \mathbf{Z}} A_i .$$

We are now in a position to prove Theorem 7. By hypothesis, there exists a surjective homomorphism

$$\mu : G \longrightarrow W .$$

Let t and a_1 be pre-images in G of the elements τ and α in W. We can find a finite set of generators of G by supplementing t and a_1 with the finite set $a_2, \ldots, a_m$, where $a_2, \ldots, a_m \in \ker \mu$. By Neumann's theorem, there is a finite presentation of G on this set of generators:

$$G = \langle\, t, a_1, \ldots, a_m \; ; \; r_1, \ldots, r_n \,\rangle .$$

Notice that

$$\exp_t r_k = 0 \qquad\qquad (k = 1, \ldots, n) \ .$$

We adopt now the notation of Theorem 6. It follows from the proof of Theorem 6 that we can decompose $H_l = \mathrm{gp}_G(t^l, a_1, \ldots, a_m)$ into an amalgamated product:

$$H_l = H_l^- \underset{C}{*} H_l^+ \ .$$

Now if we map G onto W via μ, then H_l maps onto $\mathrm{gp}_W(\tau^l, \alpha)$ and H_l^+ maps onto

$$B_l = \mathrm{gp}(\alpha_0, \ldots, \alpha_{l-1}) = A_0 \times \ldots \times A_{l-1}.$$

Observe that $a_{1,i}$ maps onto α_i $(0 \le i \le l - 1)$. We assume that l is large as compared to λ and (cf. the proof of Theorem 6) add to the presentation of H_l, the relations

$$a_{i,j} = 1 \qquad (j = 1, \ldots, m, \ 0 \le i \le \lambda - 1, \ l - \lambda \le i \le l - 1) \qquad (8).$$

It follows that the resultant quotient group, say $\overline{H}_l$, of H_l is defined by the relations (8) together with the relators

$$r_{k,i} \qquad\qquad (k = 1, \ldots, n, \ 0 \le i \le l - 1) \ .$$

The only occurences of t^l in each of these relators arise as

$$t^l a_{j,i} t^{-l} \qquad (j = 1, \ldots, m, \ 0 \le i \le \lambda - 1, \ l - \lambda \le i \le l - 1) \ .$$

This means that the addition of the relations (8) to H_l gives rise to a presentation for $\overline{H}_l$ of the form (cf. the proof of Theorem 6):

$$\begin{aligned}
\overline{H}_l = \langle \ t^l, a_{j,i} \ \ (j &= 1, \ldots, m, \ 0 \le i \le l - 1) \ ; \\
a_{j,i} &= 1 \ \ (j = 1, \ldots, m, \ 0 \le i \le \lambda - 1, \ l - \lambda \le i \le l - 1) \\
r_k(a_{1,0}, \ldots&, a_{1,\lambda}, \ldots, a_{m,0}, \ldots, a_{m,\lambda}) = 1
\end{aligned}$$

$$\vdots$$

$$\begin{aligned}
r_k(a_{1,l-\lambda-1}, \ldots, a_{1,l-1}, \ldots, a_{m,l-\lambda-1}, \ldots, a_{m,l-1}) &= 1 \\
r_k(a_{1,l-\lambda}, \ldots, a_{1,l-1}, 1, \ldots, a_{m,l-\lambda}, \ldots, a_{m,l-1}, 1) &= 1
\end{aligned}$$

$$\vdots$$

$$r_k(a_{1,l-1}, 1, \ldots, 1, \ldots, a_{m,l-1}, 1, \ldots, 1) = 1 \ \rangle \ .$$

Thus t^l does not occur in any of these relators. So

$$\overline{H}_l = \langle t^l \rangle * \mathrm{gp}\big(a_{j,i} \ \ (j = 1, \ldots, m, \ 0 \le i \le l - 1)\big) \ .$$

Here $\langle t^l \rangle$ is the image of H_l^- and $\mathrm{gp}(a_{j,i} \ (j = 1, \ldots, m, \ 0 \leq i \leq l-1))$ is the image of H_l^+. We denote these groups by

$$\overline{H}_l^- \qquad \text{and} \qquad \overline{H}_l^+ \ .$$

Hence $\overline{H}_l^-$ is infinite cyclic and $\overline{H}_l^+$ maps via μ onto

$$C_l = B_l / \mathrm{gp}(a_{j,i}\mu \ (j = 1, \ldots, m, \ 0 \leq i \leq \lambda - 1, \ l - \lambda \leq i \leq l-1)) \ .$$

Now we can assume that l is large relative to λ. Since B_l is the direct product of l cyclic groups of order h, factoring out a 2λ-generator subgroup of B_l cannot affect B_l too much. Indeed, denoting the images of the α_i in C_l again by α_i, it follows from the basis theorem for finitely generated abelian groups that,

$$C_l = \mathrm{gp}(\alpha_f) \times E_l$$

where α_f is of order h $(0 \leq f \leq l - 1)$. Putting this another way, there is a homomorphism of $\overline{H}_l^+$ onto a cyclic group of order h which maps $\alpha_{1,f}$ onto the generator of this cyclic group. So $\overline{H}_l$ itself maps onto the free product of an infinite cyclic group and a cyclic group of order h. This completes the proof of the theorem. ∎

Finally I want to prove a theorem of B. Baumslag & S.J. Pride (**J. London Math. Soc. (2) 17** (1978), 425-426). I have already alluded to this theorem. It is contained in the first of a series of three related papers – the second by the same authors (**Math. Z. 167** (1979), 279-281) and the last by R. Stohr (**Math. Z. 182** (1983), 45-47). Their results all follow from the theorem that I have just proved.

I will concern myself only with the first theorem that was proved by B. Baumslag & S.J. Pride:

Theorem 8 *Let G be a group given by $m+1$ generators and n relations ($m, n < \infty$). If*
$$(m + 1) - n \geq 2,$$
then G contains a subgroup of finite index which maps onto a free group of rank two.

Theorem 8 can be deduced from Theorem 7 and the following

Lemma 5 *Let G satisfy the hypothesis of Theorem 8. Then given any prime p there exists a surjective homomorphism*

$$\mu \ : \ G \longrightarrow W = C_p \wr C_\infty \ .$$

Proof We proceed as follows: First we present G in the form

$$G = \langle\, t, a_1, \ldots, a_m \;;\; r_1, \ldots, r_n \,\rangle,$$

where as usual

$$\exp_t r_k = 0 \qquad (k = 1, \ldots, n)\ .$$

Let p be any given prime. Let

$$N = \mathrm{gp}_G(a_1, \ldots, a_m)$$

and let

$$M = N/(N' N^p)\ .$$

Here N' is the derived group of N and N^p denotes the subgroup of N generated by the p-th powers of elements of N. Now M is an abelian group of exponent p. We write M additively and turn it into a left module over the group ring $\mathbf{F}_p[t, t^{-1}]$ of the infinite cyclic group on t over the field $\mathbf{F}_p$ of p elements:

$$\left(\sum_i c_i t^i\right)(aN'N^p) = \sum_i (t^i a t^{-i})^{c_i} N' N^p\ .$$

View M as a module on m generators (the images of the a_j in M) subject to n module relations (the images of the r_k, written in $\mathbf{F}_p[t, t^{-1}]$-module form in terms of the generators of M that I have just described). Now the ring $R = \mathbf{F}_p[t, t^{-1}]$ is a principal ideal domain. So every finitely generated R-module is a direct sum of cyclic modules. Since $n < m$ one of these cyclic modules is free. Therefore M has a quotient which is free on one generator. The submodules of M corresepond exactly to the normal subgroups of G contained in N and containing $N'N^p$. So this translates into the existence of a normal subgroup L of G contained in N such that L contains $N'N^p$ with N/L, viewed as an R-module, free on one generator, say aL. This simply means that modulo L the conjugates of a under the powers of t are independent elements of the vector space N/L. In other words

$$G/L = \mathrm{gp}(aL, tL) = \mathrm{gp}(aL) \wr \mathrm{gp}(tL).$$

This completes the proof of the lemma.   ∎

Incidentally the same kinds of arguments yield also:

Theorem 9 *Let G be a group defined by a single relation and at least three generators. Then G is virtually an amalgamated product of the form*

$$H^- \underset{U}{*} H^+$$

where U is of infinite index in H^- and also in H^+ and H^+ and H^-, H^+ and U are finitely presented.

(The books by Magnus, Karrass and Solitar, and by Lyndon and Schupp, cited in Chapter 1, are a good general reference for this chapter. See also the survey article by Strebel, cited in Chapter 1.)

Chapter V
Affine algebraic sets and the representation theory of finitely generated groups

1 Background

In the next few lectures I want to develop the very beginnings of geometric representation theory and then give two applications of this theory to combinatorial group theory.

Before I begin, however, I want to give you an idea as to how this geometric representation theory comes into play.

Let us suppose that G is a given finitely generated group. We consider the set $R(G)$ of all representations of G in $SL(2, \mathbf{C})$. This set $R(G)$ carries with it the stucture of *an affine algebraic set* (see the book by Robin Hartshorne: *Algebraic Geometry*, **Graduate Texts in Mathematics 52**, Springer-Verlag, New York (1977), for a more detailed discussion of algebraic geometry). If G has sufficiently many representations, then we can find *curves* in $R(G)$. M. Culler & P.B. Shalen, *Varieties of group representations and splittings of 3-mainifolds*, **Ann. of Math. 117 (1983) 109-146.**, in an important paper, showed how such a curve of representations can be used to produce a canonical representation

$$\gamma \; : \; G \longrightarrow SL(2, F),$$

where F is a finite algebraic extension of the field $\mathbf{C}(x)$ of rational functions over $\mathbf{C}$ in a single variable. Such a field F has a *discrete valuation*. There is a theory due to Bass, Serre and Tits which shows that, under these circumstances, $SL(2, F)$ acts as a group of automorphisms of a tree that is associated to $SL(2, F)$. Hence G also acts on a tree. The Bass-Serre part of this theory (see Chapter VII) yields a decomposition of G as the *fundamental group of a graph of groups*. This means that G can be reconstituted from a carefully chosen set of its subgroups by using HNN extensions and amalgamated products, allowing for a detailed study of G. Culler & Shalen applied this approach in the case where G is the fundamental group of a 3-manifold and obtained important new results about such groups, as well as new proofs of earlier theorems of Thurston. This technique promises also to be an important tool in combinatorial group theory. Here, I want only to sketch a few of the basic ideas involved in this geometric representation theory. In Chapter VII, I will describe the Bass-Serre theory and also the way in which $SL(2, F)$ acts on a tree.

2 Some basic algebraic geometry

Let k be a fixed algebraically closed field, n a positive integer and $\mathbf{A}_k^n = \mathbf{A}^n$ the set of all n-tuples of elements of k:

$$\mathbf{A}^n = \{ (a_1, \ldots, a_n) \mid a_j \in k \} .$$

We sometimes denote $\mathbf{A}^n$ simply by $\mathbf{A}$, which is usually referred to as *affine n-space* and its elements are then referred to as *points*.

Let n again denote a positive integer and let

$$A = k[T_1, \ldots, T_n]$$

be the polynomial algebra over k in the variables $T_1, \ldots, T_n$. A (commutative) algebra B which is isomorphic to a quotient of such an A is termed *finitely generated*. It follows that the algebra B can be generated by n elements, say $t_1, \ldots, t_n$. We express this fact by writing

$$B = k[t_1, \ldots, t_n] .$$

Notice that these generators $t_1, \ldots, t_n$ are not necessarily algebraically independent. All algebras dicussed here will be commutative, associative k-algebras with a multiplicative identity 1. We shall need two theorems of Hilbert. The first of these is

Theorem 1 (The Hilbert Basis Theorem) *Let B be a finitely generated commutative k-algebra. Then every ideal of B is finitely generated (as an ideal).*

It follows that such k-algebras satisfy the ascending chain condition for ideals.

Now let S be a subset of A. We define

$$Z(S) = \{ (a_1, \ldots, a_n) \in \mathbf{A} \mid f(a_1, \ldots, a_n) = 0 \quad \text{for all} \quad f \in S \} .$$

$Z(S)$ is termed the *zero set* or *set of zeroes* of S.

Definition 1 *A subset X of $\mathbf{A}$ is termed an affine algebraic set if $X = Z(S)$ for some subset $S \subseteq A$.*

We say X is the *affine algebraic set defined by* S. Notice that if $\mathcal{A}$ is the ideal of A generated by S, then by the Hilbert Basis Theorem $\mathcal{A}$ is finitely generated, say by $f_1, \ldots, f_r$. It follows that

$$Z(S) = Z(\mathcal{A}) = Z(f_1, \ldots, f_r) .$$

Thus

Lemma 1 *Every affine algebraic set is defined by a finite set.*

Next we record some simple facts about affine algebraic sets.

Lemma 2
(i) $\emptyset$ and $\mathbf{A}$ are affine algebraic sets.
(ii) *The union of two affine algebraic sets is again an affine algebraic set. More precisely, if $\mathcal{A}_1$ and $\mathcal{A}_2$ are ideals of A, then*

$$Z(\mathcal{A}_1) \cup Z(\mathcal{A}_2) = Z(\mathcal{A}_1 \cap \mathcal{A}_2) = Z(\mathcal{A}_1 \mathcal{A}_2) .$$

(iii) *The intersection of an arbitrary number of affine algebraic sets is again an affine algebraic set. More precisely, if $\{\mathcal{A}_i \,|\, i \in I\}$ is an indexed family of ideals of A, then*

$$\bigcap_{i \in I} Z(\mathcal{A}_i) = Z\left(\bigcup_{i \in I} \mathcal{A}_i \right) .$$

(iv) *If $S_1 \subseteq S_2 \subseteq A$ then $Z(S_1) \supseteq Z(S_2)$.*

The proof of Lemma 2 is left to the reader. Only (ii) needs a little thought – if necessary one can use (iv) to help in the proof.

Lemma 2 can be used to put a topology on $\mathbf{A}$. We take the closed sets in this topology to be the affine algebraic sets. The resultant topology on $\mathbf{A}$ is termed the *Zariski topology.*

Exercises 1 *(1) Prove Lemma 2.*

(2) Observe that
(i) $S^{n-1} = \{ (a_1, \ldots, a_n) \in \mathbf{A} \mid \sum_{j=1}^{n} a_j^2 = 1 \}$
(ii) $H = \{ (x, y) \in \mathbf{A}^2 \mid xy = 1 \}$
are affine algebraic sets.

(3) Let $M = M(n, k)$ *be the set of all $n \times n$ matrices over k. Identify M with $\mathbf{A}^{n^2}$ by simply writing down the rows of each matrix one after the other.*

(4) Prove that $SL(n, k) \subseteq \mathbf{A}^{n^2}$ *and* $GL(n, k) \subseteq \mathbf{A}^{n^2+1}$ *are affine algebraic sets.*

(5) Prove that if $k^\bullet$ is the multiplicative group of k, then

$$\underbrace{k^\bullet \times \ldots \times k^\bullet}_{l} \subseteq \mathbf{A}^{l+1}$$

is an affine algebraic set. (We identify $\underbrace{k^\bullet \times \ldots \times k^\bullet}_{l}$ with the set of $l + 1$-tuples

$(a_1, \ldots, a_{l+1})$ of elements of k satisfying the condition $a_1 \ldots a_{l+1} = 1$.)

(6) Prove that if X and Y are affine algebraic sets, then so is $X \times Y$.

Notice that the Zariski topology is not a particularly nice one. For instance the closed sets in $\mathbf{A}^1$ are the finite sets and $\mathbf{A}^1$. So the topology is not even Hausdorff.

We remind the reader of the

Definition 2 *Let G be a group. Then a representation of G in $\mathrm{SL}(n, k)$ is a homomorphism*

$$\varrho : G \longrightarrow \mathrm{SL}(n, k) .$$

Notice that if G is finitely generated, say by $g_1, \ldots, g_m$, then ϱ is completely determined by its effect on $g_1, \ldots, g_m$. So if $\mathrm{R}(G, n)$ is the set of all representations of G in $\mathrm{SL}(n, k)$ we can parametrize $\mathbf{R}(G, n)$ by points in $\mathbf{A}^{n^2 m}$. More precisely we associate with ϱ the point $(\varrho(g_1), \ldots, \varrho(g_m))$:

$$\varrho \longmapsto (\varrho(g_1), \ldots, \varrho(g_m)) .$$

Lemma 3 *Let G be a group with $g_1, \ldots, g_m$ a finite set of generators of G. Then $\mathrm{R}(G, n)$ carries with it the structure of an affine algebraic set.*

Proof Let

$$G = \langle\, g_1, \ldots, g_m\,;\, r_1 = 1, \ldots \,\rangle$$

be a presentation for G on the given generators. Notice that the number of relators in this presentation need not be finite. Let $M_1, \ldots, M_m$ be $n \times n$ matrices over k. Then a point

$$u = (M_1, \ldots, M_m) \in \mathbf{A}^{n^2 m} \tag{1}$$

corresponds to a representation ϱ of G in $\mathrm{SL}(n, k)$, i.e., the mapping defined by

$$\varrho : g_l \longmapsto M_l \qquad (l = 1, \ldots, m),$$

is a representation of G in $\mathrm{SL}(n,k)$, if and only if

$$\det M_l = 1 \qquad (l = 1,\ldots,m) \quad \text{and} \quad r_h(M_1,\ldots,M_m) = 1 \qquad (h = 1,\ldots) \, .$$

Notice that if M is any $n \times n$ matrix, then $\det M$ is a polynomial in the coefficients of M. Now if $\det M = 1$ then M^{-1} can be expressed, in the usual way, in terms of the coefficients of M. In fact each of the coefficients of M^{-1} is a polynomial in the coefficients of M. Thus if we assume that $\det M_l = 1$ for $l = 1,\ldots,m$, then $r_h(M_1,\ldots,M_m)$ is a matrix whose (i,j)-entry is a polynomial $r_h(i,j)$ in the coefficients of the M_l. Consequently $r_h(M_1,\ldots,M_m) = 1$ if and only if the polynomials $r_h(i,j) = \delta_{ij}$ (the Kronecker delta). So u (given by (1)) corresponds to a representation of G in $\mathrm{SL}(n,k)$ if and only if

$$\det M_l = 1 \qquad (l = 1,\ldots,m) \quad \text{and} \quad r_h(i,j) = \delta_{ij} \qquad (h = 1,\ldots) \, . \qquad (2)$$

It follows that $\mathrm{R}(G,n)$ is the set of zeroes of a (possibly infinite) set of polynomials which arise from a set of defining relations for G, written in terms of the given generators.

Exercises 2 *(1) Suppose that G is free on $g_1,\ldots,g_m$. Then*

$$\mathrm{R}(G,n) = \mathrm{SL}(n,k)^m \, .$$

(2) Let G be a group given by a single defining relation. Then $\mathrm{R}(G,n)$ can be defined by $m + n^2$ equations, where m is the number of generators of G.

(3) If M is any square matrix, the trace $\mathrm{tr}\, M$ of M is the sum of its diagonal entries. Verify that
(i) $\mathrm{tr}(M_1 M_2) = \mathrm{tr}(M_2 M_1)$
      where M_1 is an $n \times m$ matrix and M_2 an $m \times n$ matrix.
(ii) $\mathrm{tr}(T^{-1}MT) = \mathrm{tr}\, M$ for $M \in M(n,k)$ and $T \in \mathrm{GL}(n,k)$.
(iii) if $\det(tI - M) = f(t)$ is the characteristic polynomial of $M \in M(n,k)$ and if

$$f(t) = t^n + c_{n-1}t^{n-1} + \ldots + c_0,$$

then $c_0 = (-1)^n \det M$ and $c_{n-1} = -\mathrm{tr}\, M$.
(iv) $M \in \mathrm{SL}(2,\mathbf{C})$ is of order $e > 2$ if and only if

$$\mathrm{tr}\, M = \omega + \omega^{-1},$$

where ω is a primitive e-th root of 1.
(Hint: Use Jordan normal forms and note that if

$$\lambda = \omega + \omega^{-1},$$

then the roots of the equation $x^2 - \lambda x + 1 = 0$ are ω and ω^{-1}.)

(4) Suppose that G is a group defined by a single defining relator which is a proper power greater than 2:

$$G = \langle \, g_1, \ldots, g_m \, ; \, \big(r(g_1, \ldots, g_m)\big)^e = 1 \, \rangle \,.$$

Prove that $R(G, 2)$ can be defined by $m + 1$ equations.

On the face of it, the affine algebraic set $R(G, n)$ appears to depend on the generators chosen for G. This is not the case however. In order to prove the invariance of $R(G, n)$, we need the notion of a *morphism* from one affine algebraic set to another. We will prepare ourselves for the introduction of this and other notions in the next two sections.

3   More basic algebraic geometry

Let Y be a subset of $\mathbf{A}$. Then we define

$$I(Y) = \{ \, f \in A \mid f(a_1, \ldots, a_n) = 0 \quad \text{for all} \quad (a_1, \ldots, a_n) \in Y \, \} \,.$$

$I(Y)$ is clearly an ideal of A and is termed the *ideal of Y*. In order to better understand this function I we will need the other theorem of Hilbert that I mentioned earlier.

Theorem 2 (Hilbert's Nullstellensatz) *Let k be an algebraically closed field, $\mathcal{A}$ an ideal of $A = k[T_1, \ldots, T_n]$. If $f \in A$ vanishes on $Z(\mathcal{A})$, i.e., $f \in I\big(Z(\mathcal{A})\big)$, then a positive power of f lies in $\mathcal{A}$.*

Notice that if $\mathcal{A} \neq A$, then $Z(\mathcal{A}) \neq \emptyset$ is one of the consequences of Theorem 2. For if $Z(\mathcal{A}) = \emptyset$ then the polynomial 1 vanishes on $Z(\mathcal{A})$ and hence $1 \in \mathcal{A}$, i.e., $\mathcal{A} = A$.

Recall that if B is a commutative k-algebra then an element $b \in B$ is termed *nilpotent* if $b^r = 0$ for some positive integer r. If $\mathcal{B}$ is an ideal of B then the *radical of $\mathcal{B}$*, denoted $\sqrt{\mathcal{B}}$, is defined by

$$\sqrt{\mathcal{B}} = \{ \, b \in B \mid b^r \in \mathcal{B} \quad \text{for some} \quad r > 0 \, \} \,.$$

The radical $\sqrt{\mathcal{B}}$ of $\mathcal{B}$ is again an ideal of B and $B/\sqrt{\mathcal{B}}$ has no non-zero nilpotent elements. We need two definitions for later use.

Definition 3 *An ideal $\mathcal{B}$ of a commutative k-algebra B is termed a radical ideal if $\sqrt{\mathcal{B}} = \mathcal{B}$.*

Definition 4   *A finitely generated commutative k-algebra is termed an affine algebra if it contains no non-zero nilpotent elements.*

The function I has properties corresponding to those of Z.

Lemma 4 *Let Y, Y_1, Y_2 be subsets of **A**. Then the following hold:*
(i)   If $Y_1 \subseteq Y_2$, then $I(Y_1) \supseteq I(Y_2)$.
(ii) $I(Y_1 \cup Y_2) = I(Y_1) \cap I(Y_2)$.
(iii) If $\mathcal{A}$ is any ideal of A, then $I\,Z(\mathcal{A}) = \sqrt{\mathcal{A}}$.
*(iv) $Z\,I(Y) = \overline{Y}$, the closure of Y in the Zariski topology on **A**.*

The proof of Lemma 4 is straightforward and is left to the reader.

On allying part of Lemma 2 with Lemma 4 it follows that we have proved

Lemma 5 *Let **X** be the set of affine algebraic sets in **A** and **R** the set of radical ideals of A. Then, restricting I to **X** and Z to **R**,*

$$Z\,I = 1_{\mathbf{X}}, \qquad I\,Z = 1_{\mathbf{R}}.$$

Since both Z and I are inclusion reversing it follows that the maximal ideals of A correspond, via Z, to the minimal affine algebraic sets in **A**.

Notice that the minimal affine algebraic sets in **A** are simply points. Thus if $\mathcal{M}$ is a maximal ideal in A then

$$\begin{aligned}
\mathcal{M} &= I\,Z(\mathcal{M}) = I\,\{(a_1,\ldots,a_n)\} \\
&= \text{the ideal generated by } T_1 - a_1,\ldots,T_n - a_n.
\end{aligned}$$

Thus the maximal ideals of A all have this form. Notice also that if $x = (a_1,\ldots,a_n)$ then

$$\widehat{x} : A \longrightarrow k \text{ defined by } f \longmapsto f(x)$$

is a homomorphism of A onto k with kernel $\mathcal{M}$. So **A** is, in a sense, completely determined by the maximal ideals of A. There is, as we shall see shortly, a similar result for affine algebraic sets in general.

4   Useful notions from topology

We recall first the

Definition 5 *A topological space X is termed irreducible if*
(i) $X \neq \emptyset$ and
(ii) X is not the union of two proper closed subsets.

Then the following holds:

Lemma 6 *(i)   A topological space $X \neq \emptyset$ is irreducible if and only if every non-empty open set U is dense in X, i.e., $\overline{U} = X$.*
(ii) If Y is a subspace of a topological space then Y is irreducible if and only if $\overline{Y}$ is irreducible.
(iii) The closure of a point in a topological space X is irreducible.
(iv) If $\varphi : X \longrightarrow Y$ is a continuous map and X is irreducible, then so is $\varphi(X)$.

The proof of Lemma 6 is straightforward and is left to the reader.

Now let X be a topological space. Then, by Zorn's Lemma, every irreducible subspace is contained in a maximal one. By Lemma 6 *(ii)*, a maximal irreducible subspace is closed.

Definition 6 *Let X be a topological space. Then the maximal irreducible subspaces of X are termed the irreducible components of X.*

So by Lemma 6 *(iii)* every topological space is the union of its irreducible components.

Definition 7 *A topological space X is termed Noetherian if its open sets satisfy the ascending chain condition, i.e., every properly ascending chain of open sets is finite.*

Note that X is Noetherian if and only if it satisfies the descending chain condition for closed sets.

Lemma 7 *Every affine algebraic set X is Noetherian.*

Proof Every properly descending chain of closed subspaces of X gives rise, via I, to a properly ascending chain of ideals in A. By the Hilbert Basis Theorem this latter chain is always finite.   ■

Lemma 8 *Let X be a non-empty Noetherian topological space. Then X has only finitely many irreducible components, say $X_1, \ldots, X_m$ and*

$$X = X_1 \cup \ldots \cup X_m .$$

Proof We already know that every topological space is the union of its irreducible components (which we also know are closed). In order to prove $m < \infty$, let S be the set of all closed subspaces of X which are the union of finitely many irreducible subspaces of X. If $X \notin S$ let Y be a minimal closed non-empty subspace of X which is not in S. Then Y is certainly not irreducible. Hence $Y = Y_1 \cup Y_2$, where Y_1, Y_2 are proper closed subspaces of Y. By the minimality of Y, each of Y_1, Y_2 can be written as the union of finitely many irreducible subspaces of X. Hence so can Y. This contradiction proves the lemma. ∎

It is time to identify the irreducible affine algebraic sets.

Lemma 9 *An affine algebraic set X in $\mathbf{A}$ is irreducible if and only if $I(X)$ is a prime ideal in A.*

Proof Suppose $I(X)$ is prime and that $X = X_1 \cup X_2$ is a decomposition of $X \neq \emptyset$ into two proper closed subsets. Then

$$I(X) = I(X_1) \cap I(X_2) .$$

But $I(X)$ is a prime ideal in A. So either $I(X_1) \subseteq I(X)$ or $I(X_2) \subseteq I(X)$. So, applying Z, we find either $X_1 = X$ or $X_2 = X$, a contradiction.

On the other hand suppose that X is irreducible. We want to prove that $I(X)$ is prime. $I(X) \neq A$ since $X \neq \emptyset$. Suppose that

$$f_1 f_2 \in I(X) \qquad (f_1, f_2 \in A) .$$

Then

$$X = Z I(X) \subseteq Z(f_1 f_2) = Z(f_1) \cup Z(f_2) .$$

Hence

$$X = \left(X \cap Z(f_1) \right) \cup \left(X \cap Z(f_2) \right)$$

is a decomposition of X into two closed sets. By irreducibility, either

$$X \subseteq Z(f_1) \quad \text{or} \quad X \subseteq Z(f_2).$$

It follows that either

$$f_1 \in I(X) \quad \text{or} \quad f_2 \in I(X) .$$

Hence $I(X)$ is prime as claimed. ∎

Exercises 3 *(1) Suppose the topological space X has only finitely many irreducible components, say $X_1, \ldots, X_m$. Then*

$$X = X_1 \cup \ldots \cup X_m .$$

Prove that if

$$X = Y_1 \cup \ldots \cup Y_h \qquad (h < \infty),$$

where the Y_i are closed irreducible sets and $Y_i \not\subseteq Y_j$ if $i \neq j$, then $h = m$ and the Y's can be renumbered so that $Y_i = X_i$, for $i = 1, \ldots, m$.

(2) Prove that if $\mathcal{B}$ is any radical ideal in any finitely generated commutative k-algebra B then
(i) there are only finitely many minimal prime ideals containing $\mathcal{B}$;
(ii) $\mathcal{B}$ is the intersection of these minimal prime ideals.
(Hint: Consider first the case where $B = k[T_1, \ldots, T_n]$ and then use the third isomorphism theorem.)

We come now to an important definition.

Definition 8 *Let X be a non-empty topological space. Term a series*

$$X_0 \subset X_1 \subset \ldots \subset X_m$$

of distinct irreducible closed subspaces of X an irreducible chain of length m. Then we define the dimension of X, denoted $\dim X$, by

$$\dim X = supremum \ of \ lengths \ of \ irreducible \ chains \ in \ X \ .$$

We note some simple consequences of the definition.

Lemma 10 *Let X be a finite dimensional irreducible topological space. Then the following hold:*
(i) If $\dim X = 0$ then the closure of every 1-point set is X. So if X is T_1, i.e., every 1-point set is closed then X is a space with a single point.
(ii) If $\dim X > 0$ and Y is a closed proper subspace of X, then $\dim Y < \dim X$.

Proof We verify *(ii)*, which is almost obvious. Indeed let

$$Y_0 \subset Y_1 \subset \ldots \subset Y_m$$

be an irreducible chain in Y of length m. Since Y is closed this is an irreducible chain in X. But then

$$Y_0 \subset Y_1 \subset \ldots \subset Y_m \subset X$$

is an irreducible chain in X of length $m + 1$. This proves that $\dim Y < \infty$ and also that $\dim Y < \dim X$. ∎

Exercise 4 *If X is a topological space of finite dimension m, prove that every subspace Y of X is of finite dimension at most m.*

5 Morphisms

Let X be an affine algebraic set contained in $\mathbf{A}$.

Definition 9 *A map*

$$\mu \ : \ X \longrightarrow k$$

is termed a polynomial function if there exists a polynomial $f = f(T_1,\ldots,T_n) \in A$ *such that*

$$\mu(a_1,\ldots,a_n) = f(a_1,\ldots,a_n) \qquad \big((a_1,\ldots,a_n) \in X\big) \ .$$

Thus the polynomial functions on X are simply the polynomials in A restricted to X.

The set of such functions is denoted by $k[X]$ and becomes a k-algebra using scalar multiplication and coordinate-wise addition and multiplication of functions.

Definition 10 $k[X]$ *is termed the coordinate algebra of* X.

Notice that for each $f \in A$ we have the polynomial function f_X, which is f restricted to X. The map

$$f \longmapsto f_X \qquad (f \in A)$$

is a homomorphism of A onto $k[X]$. The kernel of this homomorphism is simply $I(X)$. So

Lemma 11 $k[X] \cong A/I(X)$.

Thus the coordinate algebras of affine algebraic sets are affine algebras since $I(X)$ is a radical ideal of A. Let t_i be the function $T_i \,|\, X$. Then

$$k[X] = k[t_1,\ldots,t_n] \ .$$

The functions t_i are termed the i-th *coordinate functions of* X $(i = 1,\ldots,n)$. It follows immediately from the definitions that

$$t_i(a_1,\ldots,a_n) = a_i \qquad (i = 1,\ldots,n) \ .$$

Exercises 5 *Verify the following:*
(1) $k[\mathbf{A}] = A$.
(2) If X is a 1-point set, $k[X] \cong k$.

(3) Let X and Y be affine algebraic sets. Prove, remembering that $X \times Y$ is an affine algebraic set, that

$$k[X \times Y] \cong k[X] \underset{k}{\otimes} k[Y] .$$

By Lemma 9 an affine algebraic set X is irreducible if and only if $I(X)$ is prime. We reformulate this remark as

Lemma 12 *An affine algebraic set X is irreducible if and only if $k[X]$ is a domain.*

Definition 11   *Let $X \subseteq \mathbf{A}^m$, $Y \subseteq \mathbf{A}^n$ be affine algebraic sets. Then the map*

$$\varphi \; : \; X \longrightarrow Y$$

is termed a morphism from X to Y if there exist $f_1, \ldots, f_n \in k[X]$ such that

$$\varphi(a_1, \ldots, a_m) = (f_1(a_1, \ldots, a_m), \ldots, f_n(a_1, \ldots, a_m))$$

for all $(a_1, \ldots, a_m) \in X$.

Definition 12   *We term two affine algebraic sets X and Y isomorphic, if there exist morphisms*

$$\varphi \; : \; X \longrightarrow Y , \qquad \gamma \; : \; Y \longrightarrow X,$$

such that

$$\gamma\varphi = 1_Y , \qquad \varphi\gamma = 1_X .$$

Note   A morphism $\varphi : X \longrightarrow Y$ from one affine algebraic set to another can be viewed as a (possibly singular) polynomial change of coordinates and $\varphi(X)$ then is "X re-expressed in terms of these new coordinates".

Lemma 13   *Let $X \subseteq \mathbf{A}^m$, $Y \subseteq \mathbf{A}^n$ be affine algebraic sets. A morphism $\varphi : X \longrightarrow Y$ is a continuous map in the Zariski topology.*

Proof   Suppose that $\varphi(a_1, \ldots, a_m) = (f_1(a_1, \ldots, a_m), \ldots, f_n(a_1, \ldots, a_m))$, where each f_i is a polynomial in m variables. We have only to prove that if Z is a closed set in Y, then $\varphi^{-1}(Z)$ is closed in X. Now if $g_1, \ldots, g_h$ are polynomials which define Z then the polynomials

$$g_i(f_1, \ldots, f_n) \qquad (i = 1, \ldots, h)$$

define $\varphi^{-1}(Z)$.   ∎

We are now in a position to prove that the set of all representations of a finitely generated group into $SL(2, \mathbf{Z})$ can be thought of as an affine algebraic set.

Lemma 14 *Let G be a given finitely generated group, d a positive integer. Suppose that*

$$G = \mathrm{gp}(g_1, \ldots, g_m) \qquad (m < \infty)$$

and also that

$$G = \mathrm{gp}(h_1, \ldots, h_n) \qquad (n < \infty) \, .$$

Then the respective affine algebraic sets $R(G, d)$, $\widetilde{R}(G, d)$ associated with G are isomorphic.

Proof Notice that there exist words w_i and v_j such that

$$h_i = w_i(g_1, \ldots, g_m) \qquad (i = 1, \ldots, n)$$

and

$$g_j = v_j(h_1, \ldots, h_n) \qquad (j = 1, \ldots, m) \, .$$

$R(G, d)$ is parametrized in $\mathbf{A}^{d^2 m}$ by the affine algebraic set

$$X = \{ \, (\varrho(g_1), \ldots, \varrho(g_m)) \mid \varrho : G \longrightarrow SL(d, k) \text{ is a representation of } G \, \}$$

and $\widetilde{R}(G, d)$ is parametrized in $\mathbf{A}^{d^2 n}$ by the affine algebraic set

$$Y = \{ \, (\varrho(h_1), \ldots, \varrho(h_n)) \mid \varrho : G \longrightarrow SL(d, k) \text{ is a representation of } G \, \} \, .$$

Then $w_1, \ldots, w_n$ define a morphism

$$\varphi \, : \, X \longrightarrow Y$$

by

$$(\varrho(g_1), \ldots, \varrho(g_m)) \longmapsto (w_1(\varrho(g_1), \ldots, \varrho(g_m)), \ldots, w_n(\varrho(g_1), \ldots, \varrho(g_m)))$$
$$(= (\varrho(h_1), \ldots, \varrho(h_n)))$$

and a similar remark holds for $v_1, \ldots, v_m$. These maps are clearly inverses and so, using the obvious notation,

$$X \cong Y$$

as claimed. ∎

The affine algebraic sets, together with the morphisms between them, form a category, the category of affine algebraic sets, which we denote for the moment by $\mathcal{C}$. Similarly, the affine k-algebras, together with the k-algebra homomorphisms between them, form a second category, the category of affine k-algebras, which we denote, again for the moment, by $\mathcal{D}$. The following theorem then holds.

Theorem 3 *The categories $\mathcal{C}$ and $\mathcal{D}$ are equivalent.*

Proof The equivalence between $\mathcal{C}$ and $\mathcal{D}$ is defined by the following contravariant functor $\mathcal{F} : \mathcal{C} \longrightarrow \mathcal{D}$. First we define $\mathcal{F}$ on objects:

$$\mathcal{F}(X) = k[X] \ .$$

Second on morphisms: if $\varphi : X \longrightarrow Y$, then

$$\mathcal{F}(\varphi) = \varphi^{\star} \ : \ k[Y] \longrightarrow k[X]$$

where

$$\varphi^{\star}(f) = f\varphi \ .$$

It is easy to check that $\varphi^{\star}$ is a k-homomorphism of k-algebras and that

$$\mathcal{F}(\psi\varphi) = \mathcal{F}(\varphi)\mathcal{F}(\psi) \ , \qquad \mathcal{F}(1) = 1 \ .$$

The corresponding functor $\mathcal{G}$ from $\mathcal{D}$ to $\mathcal{C}$ involves choosing for each affine k-algebra B a finite set $b_1, \ldots, b_m$ of generators:

$$B = k[b_1, \ldots, b_m] \ .$$

This choice is made, independently, for each affine k-algebra. Thus different choices may well be made for isomorphic k-algebras, but once a choice is made, we stick to it.

Now let

$$\sigma \ : \ k[T_1, \ldots, T_m] \longrightarrow B$$

be the homomorphism defined by

$$\sigma \ : \ T_i \longrightarrow b_i \qquad (i = 1, \ldots, m) \ .$$

Let

$$\mathcal{B} = \ker \sigma \ .$$

Then we define our functor $\mathcal{G} : \mathcal{D} \longrightarrow \mathcal{C}$ on objects as follows:

$$\mathcal{G}(B) = X$$

where

$$X = Z(\mathcal{B}) \qquad (\subseteq \mathbf{A}^m) \ .$$

Notice that
$$k[X] \cong k[T_1,\ldots,T_m]/\mathcal{B} \cong B .$$

An element $b \in B$ can then be viewed as a polynomial function on X; indeed, using the above isomorphism, b_i defines the i-th coordinate function on X.

Now we need to define $\mathcal{G}$ on k-algebra homomorphisms. Let then
$$\vartheta \;:\; B \longrightarrow C$$

be a homomorphism from one affine k-algebra into another. Both B and C come equipped with finite sets of generators:
$$B = k[b_1,\ldots,b_m] , \qquad C = k[c_1,\ldots,c_n] .$$

So
$$\vartheta(b_i) = w_i(c_1,\ldots,c_n) \qquad (i = 1,\ldots,m) ,$$

where w_i is a polynomial over k in n-variables. Notice that
$$\mathcal{G}(B) = X \subseteq \mathbf{A}^m , \qquad \mathcal{G}(C) = Y \subseteq \mathbf{A}^n .$$

Define
$$\mathcal{G}(\vartheta) \;:\; Y \longrightarrow X$$

by
$$\mathcal{G}(\vartheta)(a_1,\ldots,a_n) = \big(w_1(a_1,\ldots,a_n),\ldots,w_m(a_1,\ldots,a_n) \big) .$$

It follows that $\mathcal{G}$ is a contravariant functor from $\mathcal{D}$ to $\mathcal{C}$ and that
$$\mathcal{G}\mathcal{F} \simeq 1_{\mathcal{C}} , \qquad \mathcal{F}\mathcal{G} \simeq 1_{\mathcal{D}} . \qquad \blacksquare$$

There are a couple of consequences of the proof that I want to draw attention to.

Corollary 1 *Let X and Y be affine algebraic sets and let $\varphi \;:\; X \longrightarrow Y$ be a dominant morphism, i.e., $\varphi(X)$ is dense in Y. Then*
$$\varphi^* \;:\; k[Y] \longrightarrow k[X]$$

is a monomorphism.

Corollary 2 *Let $\vartheta \;:\; B \longrightarrow C$ be a monomorphism from the affine k-algebra B to the affine k-algebra C. Then*
$$\mathcal{G}(\vartheta) \;:\; \mathcal{G}(C) \longrightarrow \mathcal{G}(B)$$

is a dominant morphism.

We shall make frequent use of Corollary 2, especially in the case where ϑ is actually an inclusion. Indeed we note here, for later ease of exposition, the following special case of Corollary 2, which follows immediately from what has already been noted.

Corollary 3 *Let $\vartheta : B \hookrightarrow C$ be an inclusion of affine k-algebras and suppose that*
$$B = k[s_1, \ldots, s_m] , \qquad C = k[s_1, \ldots, s_m, t_1, \ldots, t_n] .$$
Suppose that $\mathcal{G}(B) = X \subseteq \mathbf{A}^m$, $\mathcal{G}(C) = Y \subseteq \mathbf{A}^{m+n}$. If
$$X_0 = \{ (z_1, \ldots, z_m) \mid \text{ there exist } a_1, \ldots, a_n \in k$$
$$\text{with the property that } (z_1, \ldots, z_m, a_1, \ldots, a_n) \in Y\},$$
then
$$X = \overline{X_0} .$$

If $\varphi : X \longrightarrow Y$ is a dominant morphism of affine algebraic sets, we sometimes refer to Y as a *quotient affine algebraic set*. Our objective is to construct such quotients, using Corollary 2.

Exercise 6 *Verify the conclusion of Corollary 3 when ϑ is the canonical inclusion of $k[T]$ in $k[T, T^{-1}]$:*
$$\vartheta : k[T] \hookrightarrow k[T, T^{-1}] .$$

6   Dimension

We begin with

Definition 13 *An affine algebraic set X is termed an affine variety if $I(X)$ is a prime ideal.*

Notice that X is an affine variety if and only if $k[X]$ is a domain.

Exercises 7 *(1) Let $f \in k[T_1, \ldots, T_n] = A$, the polynomial algebra in n variables. Then A is a unique factorization domain. Suppose $f \neq 0$. Then $Z(f)$ is an affine variety if and only if f is prime.*

(2) Now let $n = d^2$ and label the T_l $(l = 1, \ldots, n)$ as T_{ij} $(i, j = 1, \ldots, d)$. Let
$$f = \det(T_{ij}) .$$
So f is a polynomial of degree d. Prove that $f - 1$ is prime and hence that $\mathrm{SL}(d, k)$ is an affine variety.

(3) Prove that if X and Y are affine varieties, then so is $X \times Y$.

If $X \subseteq \mathbf{A}$ is an affine algebraic set, then $\dim X$ is, by definition, the length m of the longest chain

$$X_0 \subset X_1 \subset \ldots \subset X_m$$

of distinct irreducible closed subspaces of X. Applying the operator I yields a chain

$$\mathcal{X}_0 \supset \mathcal{X}_1 \supset \ldots \supset \mathcal{X}_m \tag{3}$$

of prime ideals of $k[X]$. So $\dim X$ is what is termed the *Krull dimension* of the affine k-algebra $k[X]$.

We take for granted some facts about Krull dimension:

Theorem 4 *Suppose that the affine k-algebra B is a domain. Then*
(i) $\dim B$ is the transcendence degree over k of the field of fractions of B; hence $\dim B < \infty$.
(ii) if $b_1, \ldots, b_q \in B$ and if $\mathcal{B}$ is a minimal prime ideal of B containing $b_1, \ldots, b_q$ – so by assumption $\mathcal{B} \neq B$ – then

$$\dim(B/\mathcal{B}) \geq \dim B - q .$$

Examples 1 *(1)* $\dim \mathbf{A}^n = \dim k[\mathbf{A}^n] = \dim k[T_1, \ldots, T_n] = n$.

(2) $\dim \operatorname{SL}(d, k) = d^2 - 1$. *Notice that*

$$k[\operatorname{SL}(d,k)] = k[T_{11}, \ldots, T_{dd}]/\big(\det(T_{ij}) - 1\big) .$$

Since the ideal generated by the polynomial $\det(T_{ij}) - 1$ is prime, Theorem 4, (ii) applies, yielding the desired dimension for $\operatorname{SL}(d, k)$.

(3) Let X and Y be affine varieties of dimension m and n respectively. Then

$$\dim (X \times Y) = m + n = \dim X + \dim Y .$$

Since X is an affine variety, $k[X]$ is a domain. We denote by $k(X)$ its field of fractions. Now $X \times Y$ is again an affine variety since

$$k[X \times Y] \cong k[X] \underset{k}{\otimes} k[Y]$$

is a domain. We observe, leaving the proof to the reader, that the field of fractions of $k[X] \underset{k}{\otimes} k[Y]$ is simply $k(X) \underset{k}{\otimes} k(Y)$. Now let $u_1, \ldots, u_m$, $v_1, \ldots, v_n$ be transcendence bases for $k(X)$ over k, $k(Y)$ over k, respectively. Then

$$u_1 \otimes 1, \ldots, u_m \otimes 1, \quad 1 \otimes v_1, \ldots, 1 \otimes v_n$$

is a transcendence basis for $k(X) \underset{k}{\otimes} k(Y)$ over k. This proves the assertion about the dimension of $X \times Y$.

(4) Let F be the free group on q free generators. Then $R(F,d)$ is an affine variety of dimension $q(d^2 - 1)$. We have already seen that

$$R(F,d) = \underbrace{SL(d,k) \times \ldots \times SL(d,k)}_{q} .$$

So

$$\dim R(F,d) = q(d^2 - 1) .$$

Note in particular that

$$\dim R(F,2) = 3q .$$

(5) Let G be a group defined by q generators and n defining relations. Then

$$\dim R(G,d) \geq (q - n)(d^2 - 1) .$$

So if $q > n$, then $\dim R(G,d) > 0$ for $d \geq 2$.

We compute $\dim R(G,d)$ by using (ii) of Theorem 4. Thus notice that

$$\dim R(G,d) = \dim k[R(G,d)] ,$$

and that if $f_1, \ldots, f_l$ define $R(G,d)$ then

$$k[R(G,d)] \cong k[\mathbf{A}^{qd^2}]/\sqrt{(f_1, \ldots, f_l)} .$$

Let B be a minimal prime ideal containing $\sqrt{(f_1, \ldots, f_l)}$. Since $R(G,d)$ always contains the trivial representation, i.e., the representation that maps every element of G to the identity matrix, $R(G,d) \neq \emptyset$. So $\sqrt{(f_1, \ldots, f_l)} \neq k[\mathbf{A}^{qd^2}]$. Hence B exists and so we need only estimate l and then apply Theorem 4, (ii). Suppose then that $\varrho \in R(G,d)$ and that in the parametrization of $R(G,d)$

$$\varrho \longmapsto (M_1, \ldots, M_q) .$$

In order to ensure ϱ is a representation of G in $SL(d,k)$ we need first to make sure that

$$\det M_i = 1 \qquad (i = 1, \ldots, q) .$$

This requires q polynomial equations. In addition we need

$$r_j(M_1, \ldots, M_q) = 1 \qquad (j = 1, \ldots, n)$$

for each of the n defining relations for G. On the face of it this entails d^2 equations ensuring that the coefficients of the matrix $r_j(M_1, \ldots, M_q)$ are either 0 or 1. Now the determinant of $r_j(M_1, \ldots, M_q)$ is 1. So we need only $d^2 - 1$ of these equations in *order to guarantee that the off-diagonal entries are* 0 *and that the diagonal entries are* 1. *Hence*

$$\dim\ R(G,d)\ \geq\ qd^2 - q - n(d^2 - 1) = (q-n)(d^2 - 1)\ .$$

(6) Let

$$G = \langle g_1,\ldots,g_q\,;\ \big(r(g_1,\ldots,g_q)\big)^e = 1\rangle$$

where r *is a non-trivial cyclically reduced* $\{g_1,\ldots,g_q\}$-*word and* $e > 2$. *Assume that there exists a representation of* G *in* SL $(2,\mathbf{C})$ *such that the image of* $r(g_1,\ldots,g_q)$ *is of order* e. *Verify that*

$$\dim\ R(G,2)\ \geq\ 3q-1\ .$$

Suppose next that $\varphi\ :\ X \longrightarrow Y$ is a morphism of affine algebraic sets.

Definition 14 *The fibres of* φ *are the closed sets* $\varphi^{-1}(y)$ $(y \in Y)$.

The following theorem is a useful tool in computing dimensions.

Theorem 5 *Let* $\varphi\ :\ X \longrightarrow Y$ *be a morphism of affine varieties. If* $y \in \varphi(X)$ *then each of the irreducible components* Z *of* $\varphi^{-1}(y)$ *has dimension at least* $\dim\ X - \dim\ Y$:

$$\dim\ Z\ \geq\ \dim\ X - \dim\ Y\ .$$

We shall have occasion to use Theorem 5 later.

7   Representations of the free group of rank two in SL$(2,\mathbf{C})$

I want now to study the affine algebraic set $R(F,2)$ of representations in SL$(2,\mathbf{C})$, of the free group F of rank 2 on a and b.

We already know that

$$R(F,2) = \mathrm{SL}(2,\mathbf{C}) \times \mathrm{SL}(2,\mathbf{C})\ .$$

So

$$\dim R(F,2) = 6\ .$$

We consider now the map

$$\varphi\ :\ R(F,2) \longrightarrow \mathbf{C}^3 \quad \text{given by} \quad \varrho \longmapsto (\mathrm{tr}\varrho(a), \mathrm{tr}\varrho(b), \mathrm{tr}\varrho(ab))\ .$$

Thinking of φ as a map on the affine algebraic set representing $R(F,2)$ it is clear that φ is a polynomial map:

$$\varphi(a_{11}, a_{12}, a_{21}, a_{22}, b_{11}, b_{12}, b_{21}, b_{22})$$
$$= (a_{11} + a_{22},\; b_{11} + b_{22},\; a_{11}b_{11} + a_{12}b_{21} + a_{21}b_{12} + a_{22}b_{22})$$

i.e., φ is a morphism from the affine algebraic set $R(F,2)$ to the affine algebraic set $\mathbf{C}^3$.

Lemma 15 φ *is onto.*

Proof Let $(z_1, z_2, z_3) \in \mathbf{C}^3$ and consider the quadratic equations

$$\lambda^2 - z_1\lambda + 1 = 0 \qquad \text{and} \qquad \mu^2 - z_2\mu + 1 = 0$$

for λ and μ. Notice that

$$\lambda + \lambda^{-1} = z_1 \qquad \text{and} \qquad \mu + \mu^{-1} = z_2 \; .$$

Put

$$A = \begin{pmatrix} \lambda & 0 \\ z & \lambda^{-1} \end{pmatrix} \qquad \text{and} \qquad B = \begin{pmatrix} \mu & 1 \\ 0 & \mu^{-1} \end{pmatrix},$$

with z still to be determined. Then

$$\operatorname{tr} A = z_1 \qquad \text{and} \qquad \operatorname{tr} B = z_2 \; .$$

Notice also that $\det A = 1 = \det B$. Now compute

$$AB = \begin{pmatrix} \lambda\mu & \lambda \\ z\mu & z + \lambda^{-1}\mu^{-1} \end{pmatrix} \; .$$

So

$$\operatorname{tr}(AB) = \lambda\mu + \lambda^{-1}\mu^{-1} + z \; .$$

Now put

$$z = z_3 - \lambda\mu - \lambda^{-1}\mu^{-1} \; .$$

Let $\varrho \in R(F,2)$ be defined by

$$\varrho(a) = A \qquad \text{and} \qquad \varrho(b) = B \; .$$

Then

$$\varphi(\varrho) = (z_1, z_2, z_3) \; .$$ ∎

We now make use of Lemma 15 to prove the following

Theorem 6 *Let*

$$G = \langle a, b\,; a^l = b^m = (ab)^n = 1 \rangle \qquad (l, m, n > 1)\ .$$

Then a is of order l, b is of order m and ab is of order n.

In order to prove Theorem 6 we need the following

Lemma 16 *Let* M *be of order* 2 *in* $\mathrm{SL}(2, \mathbf{C})$. *Then* M $= -1$.

Proof M is conjugate to a matrix

$$N = \begin{pmatrix} x & 0 \\ y & x^{-1} \end{pmatrix}\ .$$

Thus

$$N^2 = \begin{pmatrix} x^2 & 0 \\ y(x + x^{-1}) & x^{-2} \end{pmatrix}\ .$$

Hence $x = \pm 1$. If $x = 1$ we get a contradiction: $N = 1$ or N is of infinite order. So $x = -1$. But $y(x + x^{-1}) = 0$. So $y = 0$. Thus M is conjugate to -I, an element in the center of $\mathrm{SL}(2, \mathbf{C})$, i.e., M=-I and

$$\zeta\big(\mathrm{SL}(2, \mathbf{C})\big) = \{\pm I\}\ . \qquad\blacksquare$$

We are now in a position to prove Theorem 6. By Lemma 15, there exists a representation ϱ of the free group F on a and b in $\mathrm{SL}(2, \mathbf{C})$, such that

$$\mathrm{tr}\,\varrho(a) = \lambda + \lambda^{-1}, \qquad \mathrm{tr}\,\varrho(b) = \mu + \mu^{-1}, \qquad \mathrm{tr}\,\varrho(ab) = \nu + \nu^{-1}$$

where

$$\lambda,\ \mu,\ \nu$$

are primitive $2l$-th, $2m$-th and $2n$-th roots of 1. So, by *(iv)* of Exercises 2, $\varrho(a)$ is of order $2l$, $\varrho(b)$ is of order $2m$, $\varrho(ab)$ is of order $2n$. Consequently, by Lemma 16, the canonical images in $\mathrm{PSL}(2, \mathbf{C}) = \mathrm{SL}(2, \mathbf{C})/\{\pm 1\}$ of $\varrho(\alpha)$, $\varrho(\beta)$ and $\varrho(\alpha\beta)$ are of orders l, m and n respectively. This representation of F in $\mathrm{PSL}(2, \mathbf{C})$ factors through G. It follows, therefore, that the elements a, b and ab in G also have orders l, m and n, respectively. $\qquad\blacksquare$

Suppose, once again, that F is the free group on a and b. Observe that the coordinate algebra of $\mathbf{A} = \mathbf{C}^3$ is $\mathbf{C}[x, y, z]$, the polynomial algebra over $\mathbf{C}$ in three independent variables. Indeed we can take x, y, z to be the coordinate functions

$$x(z_1, z_2, z_3) = z_1, \qquad y(z_1, z_2, z_3) = z_2, \qquad z(z_1, z_2, z_3) = z_3\ .$$

Now, by Lemma 15, the map

$$\varphi \; : \; R(F, 2) \longrightarrow \mathbf{A}$$

is onto. Hence the homomorphism

$$\varphi^* \; : \; \mathbf{C}[\mathbf{A}] \longrightarrow \mathbf{C}[R(F, 2)]$$

is a monomorphism. Observe that

$$\varphi^*(x)(\varrho) = \operatorname{tr} \varrho(a) \; , \;\; \varphi^*(y)(\varrho) = \operatorname{tr} \varrho(b) \; , \;\; \varphi^*(z)(\varrho) = \operatorname{tr} \varrho(ab).$$

Observe also that, for each reduced word w in a and b, the image of $\mathbf{C}[\mathbf{A}]$ under φ^* contains the polynomial functions f_w where

$$f_w(\varrho) = \operatorname{tr} \varrho(w) \qquad (w \in F) \; .$$

Furthermore, $\varphi^*(x) = f_a$, $\varphi^*(y) = f_b$ and $\varphi^*(z) = f_{ab}$.

We are now in a position to formulate the following

Lemma 17
$$\varphi^*(\mathbf{C}[\mathbf{A}]) = \mathbf{C}[f_a, f_b, f_{ab}],$$

i.e., if w is any reduced word in a and b, then $f_w \in \mathbf{C}[f_a, f_b, f_{ab}]$.

What this means is that f_w is a polynomial in f_a, f_b and f_{ab}, i.e., there exists a polynomial p_w in three variables, such that

$$\operatorname{tr} \varrho(w) = p_w \big(\operatorname{tr} \varrho(a), \; \operatorname{tr} \varrho(b), \; \operatorname{tr} \varrho(ab) \big) \; .$$

In order to prove Lemma 17 we need the following "trace identities" for $SL(2, \mathbf{C})$:

Lemma 18
(i) $\operatorname{tr} I = 2$.
(ii) $\operatorname{tr}(AB) = \operatorname{tr} A \operatorname{tr} B - \operatorname{tr}(AB^{-1})$.

Proof It follows from the Cayley-Hamilton theorem that for each $B \in SL(2, \mathbf{C})$,

$$B^2 - (\operatorname{tr} B) \cdot B + I = 0 \; .$$

Hence

$$B + B^{-1} = (\operatorname{tr} B) \cdot 1$$

(which we also could have verified directly). Now

$$\begin{aligned}
\operatorname{tr}(AB) + \operatorname{tr}(AB^{-1}) &= \operatorname{tr}\left(A(B+B^{-1})\right) \\
&= \operatorname{tr}\left(A \cdot (\operatorname{tr}B) \cdot 1\right) \\
&= \operatorname{tr}A \operatorname{tr}B .
\end{aligned}$$

∎

Remarks (i) Putting A=I in Lemma 18, yields $\operatorname{tr}B^{-1} = \operatorname{tr}B$.

(ii) Part (*ii*) of Lemma 18 appears in formula (7) of R. Fricke, and F.Klein: *Vorlesungen über die Theorie der automorphen Functionen*, **Band 1; Leipzig: Teubner 1897, p. 338.**

The proof of Lemma 17 follows immediately, by induction, from Lemma 18.

Exercise 8 *Compute f_w for $w = aba^{-1}b^{-1}$.*

If we simply denote $\varphi^\star(x)$ again by x, $\varphi^\star(y)$ by y and $\varphi^\star(z)$ by z then we see that

$$f_w = p_w(x, y, z)$$

is a polynomial in three independent variables x, y, z. Now suppose that α and β are non-zero complex numbers and that

$$\alpha^2 \neq 1, \qquad \beta^2 \neq 1 .$$

Set

$$\lambda = \alpha + \alpha^{-1} \text{ and } \qquad \mu = \beta + \beta^{-1}.$$

T. Jorgensen has proved that if

$$w = a^{r_1} b^{s_1} \ldots a^{r_k} b^{s_k} \qquad (k \geq 1, \ r_i, s_i > 0),$$

then

$$q_w(z) = p_w(\lambda, \mu, z)$$

is a polynomial of degree k, and the coefficient of z^k is given by the formula

$$\prod_{i=1}^{k} \left(\frac{\alpha^{r_i} - \alpha^{-r_i}}{\alpha - \alpha^{-1}} \right) \left(\frac{\beta^{s_i} - \beta^{-s_i}}{\beta - \beta^{-1}} \right) .$$

Exercise 9 *Prove Jorgensen's formula.*

We record next the following

Lemma 19 *Let F be a free group on $a_1, \ldots, a_m$. Then the following hold:*
(i) If $w \in F$, $w \neq 1$, there exists a homomorphism

$$\varrho : F \longrightarrow \mathrm{SL}(2, \mathbf{C})$$

such that $\varrho(w) \neq 1$, i.e., F is residually a subgroup of $\mathrm{SL}(2, \mathbf{C})$.
(ii) If $w \in F$, $w \neq 1$, and if $n > 2$ is a given integer, then there exists a representation ϱ of F in $\mathrm{SL}(2, \mathbf{C})$ such that $\varrho(w)$ is of order n.

The proof of Lemma 19 is not difficult. One observes first that F can be embedded in a free group of rank two and then appeals to Jorgensen's formula. The details are left to the reader as an exercise.

The following corollary of Lemma 19 is a special case of a theorem of Magnus, Karrass and Solitar.

Corollary 1 *Suppose that*

$$G = \langle a_1, \ldots, a_m \, ; \, w^n = 1 \rangle \qquad (n > 1)$$

is a group with a single defining relation, where $w \neq 1$ in the free group on $a_1, \ldots, a_m$. Then $w \neq 1$ in G; indeed w is of order n in G.

The proof of Corollary 1 of Lemma 19 is analogous to the proof of Theorem 6 and is left to the reader.

Lemma 19 can be used to prove some variations of the following celebrated theorem of W. Magnus:

Theorem 7 (W. Magnus 1932) *Let*

$$G = \langle a_1, \ldots, a_m \, ; \, r = 1 \rangle$$

be a group defined by a single relation. Suppose r is cyclically reduced and involves the generator a_1. Then

$$\mathrm{gp}(a_2, \ldots, a_m)$$

is a free subgroup of G freely generated by $a_2, \ldots, a_m$.

Theorem 7 is referred to as the Freiheitssatz.

8 Affine algebraic sets of characters

We have seen how knowledge of the affine algebraic set $R(G, 2)$ of representations of a free group G can be used to obtain some non-trivial results. We did not actually make use of $R(G, 2)$. In fact we made use of a quotient variety variety $\mathbf{C}^3$ of $R(G, 2)$. Our next objective is to define such quotient varieties in general (see the paper by Culler and Shalen, cited above). Before going into the details, I want to record a purely group-theoretic result which is proved by making use of this method. In order to describe this result, I need to introduce a definition.

Definition 15 *Let $\mathcal{P}$ denote the finite presentation*

$$\langle\, x_1, \ldots, x_m\,;\, r_1, \ldots, r_n\,\rangle$$

of a group G. Then we define the deficiency def$(\mathcal{P})$ of $\mathcal{P}$ by def$(\mathcal{P}) = m - n$.

Let us denote the subgroup of a group G generated by the set $\{g^2 \mid g \in G\}$ by G^2. Then G/G^2 is an abelian group all of whose elements have order dividing 2. Hence it can be viewed as a vector space over the field of two elements. We denote the dimension of this vector space by $\dim(G/G^2)$.

The following unpublished theorem, which is due to G. Baumslag and P.B. Shalen, can be proved by using the quotient variety approach alluded to above.

Theorem 8 *Let G be a group given by a finite presentation of deficiency $\operatorname{def} G = \delta$. Suppose that*

$$3\delta - 3 \;>\; \dim(G/G^2)\,.$$

Then G is an amalgamated product where the amalgamated subgroup is of infinite index in one factor and of index at least two in the other.

As a consequence of Theorem 8 we find

Corollary 5 *Let G be a group defined by a single defining relation with at least four generators. Then G is an amalgamated product of the kind described in Theorem 8.*

Recently, in fact after these lectures had been given, Baumslag and Shalen proved that any finitely presented group, with a presentation of deficiency at least 2, can be decomposed into an amalgamated product of two groups, where the amalgamated subgroup is of index at least 2 in one factor and of index at least 3 in the other (G. Baumslag and P.B. Shalen, *Amalgamated products and finitely presented groups*, **Comment. Math. Helvetici 65 (1990) 243–254**).

Now let G be a finitely generated group. We define the quotient $X(G, d)$ of $R(G, d)$, introduced by Culler and Shalen, termed the *affine algebraic set of characters of G in* $\mathrm{SL}(d, k)$ mentioned above.

Definition 16 *Let ϱ, σ be representations of G in $\mathrm{SL}(d,k)$. Then we term ϱ and σ equivalent if there exists a matrix $\mathrm{T} \in \mathrm{SL}(d,k)$ such that*

$$\varrho(g) = \mathrm{T}\sigma(g)\mathrm{T}^{-1} \qquad \text{(for all } g \in G) .$$

In general it is not possible to parametrize the equivalence classes of representations of G by the points of an affine algebraic set. However there is a parametrisation of the semi-simple representations. We need to recall some related definitions and notions. To this end, let G be a group, k a field, V a finite dimensional vector space over k, $\mathrm{SL}(V)$ the group of all invertible linear transformations of V of determinant 1 and ϱ a representation of G in V, i.e., a homomorphism from G into $\mathrm{SL}(V)$. We say V *affords the representation* ϱ. If now $k[G]$ denotes the group algebra of G over k, then a $k[G]$-module V can be viewed as a vector space V together with a representation ϱ of G in V. Two representations ϱ and σ of G in V and W are *equivalent* if the corresponding $k[G]$-modules V and W are isomorphic. A representation ϱ of G in V is termed *irreducible* if V is a simple $k[G]$-module, i.e., has no non-trivial submodules and $V \neq 0$. Then, by Schur's Lemma, the k-algebra $\mathrm{End}\,V$ is a division ring. Since k is algebraically closed it follows that $\mathrm{End}\,V$ is simply k. V is termed *semi-simple* if it is a direct sum of simple $k[G]$-modules, in which case the corresponding representation is termed *semi-simple*.

Now suppose ϱ is a representation of G in V. Then think of V as a $k[G]$-module and let

$$0 = V_0 < \ldots < V_l = V$$

be a composition series for V. Put

$$W_i = V_i/V_{i-1} \qquad (i = 1, \ldots, l) .$$

By the Jordan-Hölder theorem the $k[G]$-modules W_i are unique up to isomorphism. Put

$$W = W_1 \oplus \ldots \oplus W_l .$$

Then W is again a $k[G]$-module, indeed a semi-simple $k[G]$-module. Let ϱ^{ss} denote the underlying representation of G in W. This representation ϱ^{ss} is unique, up to equivalence.

Suppose next that, in addition, G is finitely generated and that V is a finite dimensional vector space of dimension d over k. If we fix a basis for V, we have already seen how to view the set $\mathrm{R}(G,d)$ of all representations of G in V as an affine algebraic set. As before, then, we have the coordinate algebra $k[\mathrm{R}(G,d)]$ of $\mathrm{R}(G,d)$. For each $g \in G$ define

$$\widehat{g} \; : \; \mathrm{R}(G,d) \longrightarrow \mathrm{SL}(V)$$

by

$$\widehat{g}(\varrho) = \varrho(g) \ .$$

Since we have fixed a basis for V, we can view $\mathrm{SL}(V)$ in terms of this basis. Then $\widehat{g}$ is readily seen to be a morphism of affine algebraic sets, i.e., it is defined, as usual, by polynomial functions. Next we define, for $i = 0, \ldots, d-1$ and $g \in G$, the functions

$$f_g^i \ : \ \mathrm{R}(G,d) \longrightarrow k$$

as follows:
$$f_g^i(\varrho) = \pm \ \text{coefficient of the degree} \ i \ \text{term in the}$$
$$\text{characteristic polynomial of} \ \varrho(g) \qquad ,$$

where here the $\pm$ is chosen so as to ensure that $f_g^i(\varrho)$ has a positive sign. The remark above about $\widehat{g}$ shows that

$$f_g^i \ \in \ k[\mathrm{R}(G,d)] \qquad (i = 0, \ldots, d-1, \ g \in G) \ . \qquad (4)$$

Let C be the k-subalgebra of $k[\mathrm{R}(G,d)]$ generated by the functions f_g^i, where g here ranges over the elements of G and $i = 0, \ldots, d-1$. The following theorem is due to Procesi.

Theorem 9 C *is an affine k-algebra.*

Let $\mathrm{X}(G,d)$ be the affine algebraic set defined by C, via the categorical equivalence between affine algebraic sets and affine k-algebras, and let

$$p \ : \ \mathrm{R}(G,d) \longrightarrow \mathrm{X}(G,d)$$

be the canonical projection of $\mathrm{R}(G,d)$ to $\mathrm{X}(G,d)$ that corresponds to the inclusion $C \hookrightarrow k[\mathrm{R}(G,d)]$.

Definition 17 $\mathrm{X}(G,d)$ *is termed the affine algebraic set of characters of G in* $\mathrm{SL}(V)$.

The nature of $\mathrm{X}(G,d)$ is clarified by the next theorem, which is also due to Procesi.

Theorem 10
(i) $p \ : \ \mathrm{R}(G,d) \longrightarrow \mathrm{X}(G,d)$ *is onto.*
(ii) $p(\varrho) = p(\varrho^{ss})$.
(iii) If ϱ and σ are semi-simple representations of G in V then $p(\varrho) = p(\sigma)$ if and only if ϱ and σ are equivalent.
(iv) If ϱ is irreducible, then $p^{-1}(p(\varrho))$ is the equivalence class of representations of G in V containing ϱ.

Example 2 *Suppose that $d = 2$. Then the functions f_g^i $(i = 0, 1)$ are particularly easy to describe:*

$$f_g^0(\varrho) = \det \varrho(g) = 1 \ ;$$
$$f_g^1(\varrho) = \operatorname{tr} \varrho(g) \ .$$

Let

$$\chi_\varrho \ : \ G \longrightarrow k \quad \text{be defined by} \quad g \longmapsto \operatorname{tr} \varrho(g) \ .$$

We term χ_ϱ the *character* of ϱ. The affine k-algebra C is generated by the functions

$$f_{g_1}^1, \ldots, f_{g_l}^1 \qquad (l < \infty)$$

for some choice of elements $g_1, \ldots, g_l \in G$ and

$$p(\varrho) = \left(\ \operatorname{tr} \varrho(g_1), \ldots, \operatorname{tr} \varrho(g_l)\ \right) \ .$$

It follows from Theorem 10 that if ϱ and σ are semi-simple representations of G, then $\chi_\varrho = \chi_\sigma$ if and only if ϱ and σ are equivalent. So $X(G, d)$ parametrizes the characters of the semi-simple representations of G.

Exercise 10 *Let G be a finitely generated group. Suppose that G has an irreducible representation in V, a vector space of dimension d. If G has a finite presentation on m generators and n defining relations, prove that*

$$\dim X(G, d) \ \geq \ (m - n)(d^2 - 1) \ - \ (d^2 - 1)$$

If G is free of rank m, deduce that

$$\dim X(G, d) \ \geq \ (m - 1)(d^2 - 1) \ .$$

Chapter VI
Generalized free products and
HNN extensions

1 Applications

Recall that if a group G is a generalised free product

$$G = A \underset{H}{*} B,$$

then

(i) $G = \mathrm{gp}(A \cup B)$, where A and B are subgroups of G;
(ii) $A \cap B = H$;
(iii) every "strictly alternating" $A \cup B$-product

$$x_1 \ldots x_n \neq 1 \qquad (n > 0).$$

Here, as usual, the product $x_1 \ldots x_n$ is termed strictly alternating if each $x_i \in (A-H) \cup (B-H)$ and if $x_i \in A$ then $x_{i+1} \notin A$ and if $x_i \in B$ then $x_{i+1} \notin B$ $(i = 1, \ldots, n-1)$.

The very definition of such an amalgamated product ensures that G has the following universal mapping property: for every group X and every pair of homomorphisms

$$\alpha \; : \; A \longrightarrow X \,, \qquad \beta \; : \; B \longrightarrow X$$

such that

$$\alpha \,|\, H = \beta \,|\, H,$$

there exists a unique homomorphism

$$\mu \; : \; G \longrightarrow X$$

which agrees with α on A and β on B.

In this chapter I want to give some examples of the way in which generalized free products can be used.

To begin with we give Graham Higman's example of a finitely generated non-hopfian group (G.Higman : *A finitely related group with group with an isomorphic proper factor group*, **J. London Math. Soc. 26**, 59-61 (1951)).

Theorem 1   *There exists a finitely presented group G which is isomorphic to one of its proper factor groups.*

Proof   Let
$$A = \langle a, s; a^s = a^2 \rangle , \qquad B = \langle b, t; b^t = b^2 \rangle .$$

Recall that A and B are simply semidirect products of the dyadic fractions, i.e., the subgroup of $\mathbf{Q}$ consisting of all rational numbers of the form $l/2^m$, by an infinite cyclic group where the infinite cyclic group acts by multiplication by 2. In particular, a and b are of infinite order. So we can form the amalgamated product

$$G = \{ A * B; a = b \}$$

(using the obvious notation). Here the amalgamated subgroups are

$$H = \mathrm{gp}(a) \qquad \text{and} \qquad K = \mathrm{gp}(b) .$$

Let
$$\alpha : A \longrightarrow G$$

be defined by
$$\alpha : a \longmapsto a^2 , \; s \longmapsto s$$

(i.e., α is conjugation by s followed by the inclusion of A into G). Similarly define
$$\beta : B \longrightarrow G$$

by
$$\beta : b \longmapsto b^2 , \; t \longmapsto t .$$

Notice that α and β agree on H:

$$\alpha : a \longmapsto a^2 , \; \beta : a(= b) \longmapsto a^2 .$$

So they can simultaneously be extended to a homomorphism

$$\mu : G \longrightarrow G .$$

Since $G\mu \ni a^2, s, t$, it follows that

$$G\mu \ni a .$$

Thus

$$G\mu = G \ .$$

Now consider the element

$$g = \underbrace{sas^{-1}}\,\underbrace{tb^{-1}t^{-1}} \ .$$

If we gather sas^{-1} together and $tb^{-1}t^{-1}$ together, g becomes a strictly alternating $A \cup B$-product. So

$$g \neq 1 \ .$$

But observe that

$$g\mu = sa^2s^{-1}tb^{-2}t^{-1} = ab^{-1} = 1 \ .$$

So

$$G/\ker\mu \,\cong\, G$$

with $\ker\mu \neq 1$. ■

Next let me turn to a theorem of G. Higman, B.H. Neumann and Hanna Neumann proved in 1949.

Theorem 2 *Every countable group can be embedded in a 2-generator group.*

Proof Let G be a countable group. We enumerate the elements of G as an infinite sequence, using repetitions of the elements of G if needed:

$$G = \{\, g_0 = 1\,,\, g_1, g_2, \dots \,\}$$

Let now

$$U = \langle\, u, v\,\rangle\,, \qquad B = \langle\, a, b\,\rangle$$

be two free groups of rank two. Notice that the elements

$$u\,,\, vuv^{-1}\,,\, v^2uv^{-2}\,,\, \dots$$

freely generate a free subgroup of U, and similarly for

$$a\,,\, bab^{-1}\,,\, b^2ab^{-2}\,,\, \dots$$

in B. Now let A be the free product of G and U:

$$A = G * U \ .$$

The elements

$$g_0u\,,\, g_1vuv^{-1}\,,\, g_2v^2uv^{-2}\,,\, \dots$$

freely generate a free subgroup of A. To see that this is so, observe that every reduced word in these elements is non-trivial since it projects onto an element of U which is different from 1. Put

$$H = \mathrm{gp}(\, g_0 u \,,\, g_1 vuv^{-1} \,,\, g_2 v^2 uv^{-2} \,,\, \dots)$$

and

$$K = \mathrm{gp}(\, a \,,\, bab^{-1} \,,\, b^2 ab^{-2} \,,\, \dots) \,.$$

Now H and K are both free of countably infinite rank. So we can form the amalgamated product

$$P = \{\, A * B \,;\, H = K \,\}$$

where the equality $H = K$ is defined by identifying the elements

$$g_i v^i uv^{-i} \qquad \text{with} \qquad b^i ab^{-i} \qquad (i = 0, 1, 2, \dots) \,.$$

So A and B can be viewed as subgroups of P and

$$g_0 u = a \,,\, g_1 vuv^{-1} = bab^{-1} \,,\, g_2 v^2 uv^{-2} = b^2 ab^{-2} \,,\, \dots$$

in P. But this means that

$$P = \mathrm{gp}(u, v, a, b) \,.$$

Hence

$$P = \mathrm{gp}(v, a, b)$$

since $g_0 = 1$ and therefore $u = a$.

Now observe that v and a freely generate a free group of rank two and a and b freely generate a free group of rank two. Form the HNN extension E with base P, associated subgroups $\mathrm{gp}(v, a)$, $\mathrm{gp}(a, b)$ and associating isomorphism

$$\varphi \,:\, v \longmapsto a \,,\, a \longmapsto b$$

and stable letter t:

$$E = \langle\, P, t \,;\, tvt^{-1} = a \,,\, tat^{-1} = b \,\rangle \,.$$

Then

$$E = \mathrm{gp}(t, v)$$

is the desired 2-generator group. ∎

2 Back to basics

We need to gather together some additional information about amalgamated products and HNN extensions.

Suppose then that
$$G = A \underset{H}{*} B$$
is an amalgamated product. Here we adopt, as before, the point of view that
$$A \leq G , \; B \leq G , \; A \cap B = H .$$

Let us choose a left transversal S of H in A and a left transversal T of H in B. Then every element $g \in G$ can be expressed in the form
$$g = u_1 \ldots u_n h \qquad (n \geq 0), \tag{1}$$
where
$$u_i \in (S \cup T) - \{1\} \qquad (i = 1, \ldots, n) , \qquad h \in H$$
and successive u's come from different transversals. We define the length $l(g)$ of g by
$$l(g) = n$$
and term (1) the *normal form* of g. The following lemma justifies these notions.

Lemma 1 *Let*
$$G = A \underset{H}{*} B .$$
Then the following hold:
(i) If $g \in G - H$ is written as a strictly alternating $A \cup B$-product
$$g = x_1 \ldots x_n \qquad (n > 0) \tag{2}$$
then n depends only on g, i.e., any two such representations for g have the same number of factors.
(ii) If S and T are left transversals of H in A and B respectively, then the normal form (1) for $g \in G$ is unique.
(iii) If g is given by (2), then $l(g) = n$.
(iv) $l(g) = 0$ if and only if $g \in H$.

The proof of Lemma 1 rests on the fact that in a generalised free product, strictly alternating products of elements are not equal to 1, and is left to the reader. Similar remarks hold also in the case of an amalgamated product with more than two factors.

Returning to the amalgamated product G of Lemma 1, let us term the strictly alternating product (2) *cyclically reduced* if either $n = 1$ or if $n > 1$, if x_1 and x_n

come from different factors A, B. Since this is a property of g itself we say g is *cyclically reduced*.

Lemma 2 *Let*
$$G = A \underset{H}{*} B .$$
Then every element of G which is cyclically reduced and of length at least two is of infinite order.

Proof Let $g \in G$ be cyclically reduced and of length at least two. We write g in strictly alternating form
$$g = x_1 \ldots x_n \qquad (n \geq 2) .$$
Then for every $m > 0$,
$$g^m = x_1 \ldots x_n x_1 \ldots x_n \ldots x_1 \ldots x_n . \qquad (3)$$
Since g is cyclically reduced, x_1 and x_n lie in different factors. It follows from (3) that g^m is also cyclically reduced and of length $mn \geq 2$. So
$$g^m \neq 1,$$
indeed g^m does not even lie in H. ∎

Corollary 1 *In an amalgamated product every element of finite order is conjugate to an element in one of the factors. Hence an amalgamated product of torsion-free groups is torsion-free.*

We prove next

Lemma 3 *Let $G = \{\prod_{i \in I}^* G_i; H\}$. Suppose that for each $i \in I$, F_i is a subgroup of G_i and that*
$$F_i \cap H = K = F_j \cap H \qquad (i, j \in I) .$$
Then
$$\mathrm{gp}\left(\bigcup_{i \in I} F_i\right) = \{\prod_{i \in I}^* F_i \; ; \; H \}.$$

Proof Every strictly alternating $\bigcup_{i \in I} F_i$-product, relative to K, is non-trivial because it is a strictly alternating $\bigcup_{i \in I} F_i$-product, relative to H. ∎

Lemma 3 is Corollary 4 of Chapter III; we have given a proof here for completeness.

It is time to turn our attention to HNN extensions, a special case of which was introduced in Chapter IV.

Definition 1 *Let $B = \langle X; R \rangle$ be a presentation of a given group B. An HNN extension E with base B, associated subgroups H_i, K_i $(i \in I)$, associating isomorphisms $\varphi_i : H_i \xrightarrow{\sim} K_i$ $(i \in I)$ and stable letters t_i $(i \in I)$ is the group given by the presentation*

$$E = \langle\, X \dot{\cup} \{t_i \,|\, i \in I\}\,;\ R \cup \{t_i h t_i^{-1} (h\varphi_i)^{-1} \,|\, h \in H_i,\, i \in I\}\,\rangle\,.$$

As we have already noted previously, groups given by generators and defining relations can be difficult to unravel. Our objective now is to show that E can be reasonably well understood by finding an isomorphic copy of E in a suitably chosen amalgamated product. To this end let

$$U = B * \langle u_i \,|\, i \in I \rangle \quad,\quad V = B * \langle v_i \,|\, i \in I \rangle\,.$$

Observe that the subgroup C of U generated by B together with the conjugates $u_i H_i u_i^{-1}$ of the H_i $(i \in I)$, is their free product:

$$C = \mathrm{gp}\big(B,\, u_i H_i u_i^{-1}\,(i \in I)\big) = B * \prod_{i \in I} u_i H_i u_i^{-1}\,.$$

Similarly, in V we find that

$$D = \mathrm{gp}\big(B,\, v_i^{-1} K_i v_i\,(i \in I)\big) = B * \prod_{i \in I} v_i^{-1} K_i v_i\,.$$

There is an obvious isomorphism

$$\varphi : C \xrightarrow{\sim} D$$

which is the identity on B and maps $u_i H_i u_i^{-1}$ onto $v_i^{-1} K_i v_i$ as prescribed by φ_i. So we can form the generalized free product

$$G = \{\, U * V\,;\ C \overset{\varphi}{=} D\,\}$$

using φ to identify C and D. We understand G well since it is an amalgamated product. This means, in particular, that B embeds into G, and if we set

$$\widetilde{t}_i = v_i u_i \qquad (i \in I),$$

then

$$\widetilde{t}_i h \widetilde{t}_i^{-1} = h\varphi_i \qquad (h \in H_i,\, i \in I)\,.$$

We claim that E is isomorphic to the subgroup $\widetilde{E}$ of G defined by

$$\widetilde{E} = \mathrm{gp}\big(B,\, \widetilde{t}_i\,(i \in I)\big)\,.$$

This is easy enough to check in a number of ways. For example, by Dyck's Theorem, we are assured of the existence of a homomorphism

$$\alpha \; : \; E \longrightarrow \tilde{E}$$

mapping B identically to B and t_i to $\tilde{t}_i$ $(i \in I)$. On the other hand, we define a homomorphism γ of G onto E by first describing its effect on U and V as follows:

$$\gamma \,|\, B = \text{id} \,, \qquad \gamma(u_i) = 1 \quad (i \in I) \,,$$

$$\gamma \,|\, B = \text{id} \,, \qquad \gamma(v_i) = t_i \quad (i \in I) \,.$$

Now γ has the same effect on C and D and so it can be continued to a homomorphism of G onto E. Notice that if we put

$$\beta = \gamma \,|\, \tilde{E},$$

then α and β are mutually inverse, as required. This then permits us to deduce what we have already assumed about HNN extensions, plus a little more. We record these deductions here as

Theorem 3 *Let*

$$E = \langle \, B \,, t_i \; (i \in I) \,; \; t_i H_i t_i^{-1} = K_i \; (i \in I) \, \rangle$$

be an HNN extension. The following hold:
(i)   the canonical homomorphism of B into E is a monomorphism;
(ii) if

$$w = b_0 t_{j_1}^{\varepsilon_1} b_1 t_{j_2}^{\varepsilon_2} \ldots t_{j_n}^{\varepsilon_n} b_n \qquad\qquad (n > 0)$$

where

$$j_1, \ldots, j_n \in I \,, \qquad b_0, \ldots, b_n \in B \,, \qquad \varepsilon_1, \ldots, \varepsilon_n \in \{1, -1\}$$

and if

$$w = 1$$

then there exists a "pinch" in w, i.e., for some m, $1 \leq m \leq n - 1$, either

$$\varepsilon_m = 1 \,, \; b_m \in H_{j_m} \qquad and \qquad j_{m+1} = j_m \,, \; \varepsilon_{m+1} = -1$$

or else

$$\varepsilon_m = -1 \,, \; b_m \in K_{j_m} \qquad and \qquad j_{m+1} = j_m \,, \; \varepsilon_{m+1} = 1 \,.$$

The point of this theorem is this: if $w =_E 1$, then there is an obvious way of reducing the number of *t-symbols* in w.

Part (i) of Theorem 3 goes back to Higman, B.H. Neumann and Hanna Neumann in 1949, while (ii) is an observation due to J.L. Britton and is usually referred to as *Britton's Lemma*. It follows immediately from our reconstruction of E as a subgroup of the generalized free product G.

Exercise 1 *The HNN extension E contains elements of finite order only if the base group B does.*

3 More applications

Theorem 4 (B.H.Neumann) *There exist continuously many non-isomorphic 2-generator groups.*

Proof Let p_i denote the i-th prime and let $C(i)$ denote the cyclic group of order p_i. For each increasing sequence $\sigma = (\sigma(i))_{i=1,2,...}$ of positive integers, define

$$A_\sigma = \bigoplus_{i=1}^{\infty} C(\sigma(i)) \, ,$$

the direct sum of the cyclic groups of order $p_{\sigma(i)}$. Notice that

$$A_\sigma \cong A_\tau \quad \text{if and only if} \quad \sigma = \tau \, ,$$

since $\sigma \neq \tau$ implies A_σ and A_τ do not have the same finite subgroups. Now embed each A_σ in a 2-generator group G_σ, using the method in Section 1. Then it follows from the information obtained in Section 2 that the elements of finite order are precisely those which are conjugate to elements in A_σ. So

$$G_\sigma \cong G_\tau \quad \text{if and only if} \quad \sigma = \tau \, .$$

Now the cardinality of such sequences $(\sigma(i))_{i=1,2,...}$ is that of the continuum. This proves Theorem 4.   ∎

Recall next, that a property $\mathcal{M}$ of finitely presented groups is termed a *Markov property*, if it is preserved under isomorphism, if there exists a finitely presented group G_1 with $\mathcal{M}$ and a finitely presented group G_2 which cannot be embedded in any finitely presented group with $\mathcal{M}$.

My objective is to give Rabin's proof of the following theorem, which is essentially contained in the work of Adian (see the book by R.C. Lyndon and Paul E. Schupp: *Combinatorial Group Theory*, **Ergebnisse der Mathematik und ihrer Grenzgebiete 89**, Springer-Verlag, Berlin-Heidelberg-New York (1977)). The proof makes use of the existence of a finitely presented group with an unsolvable word problem.

Theorem 5 (Adian, Rabin) *Let $\mathcal{M}$ be a Markov property. Then there is no algorithm which decides whether or not a group, given by an arbitrary finite presentation, has $\mathcal{M}$.*

Proof   Let U be a finitely presented group with an unsolvable word problem. Put

$$U_0 = U * G_2$$

where G_2 is the finitely presented group which cannot be embedded in any finitely presented group with $\mathcal{M}$. U_0 is finitely presented since U and G_2 are finitely presented:

$$U_0 = \langle\, x_1, \ldots, x_m \; ; \; r_1, \ldots, r_n \,\rangle\,.$$

Since U is a subgroup of U_0, U_0 has an unsolvable word problem. We will construct, for each reduced word $w = w(x_1, \ldots, x_m)$ in the generators of U_0, a finitely presented group G_w with the following property:

$$G_w \text{ has } \mathcal{M} \qquad \text{if and only if} \qquad w =_{U_0} 1.$$

This suffices to prove the theorem since U_0 has an unsolvable word problem.

The construction of G_w is carried out in stages.
First we form

$$U_1 = U_0 * \langle y_0 \rangle$$

the free product of U_0 and the infinite cyclic group on y_0. Observe that if we put

$$y_i = y_0 x_i \qquad (i = 1, \ldots, m)$$

then the y_i are all of infinite order and

$$U_1 = \mathrm{gp}(y_0, y_1, \ldots, y_m)\,.$$

Notice also that if $w \neq 1$, then $[w, y_0] \neq 1$; indeed $[w, y_0]$ is of infinite order (it is cyclically reduced and of length at least two). Now form an HNN extension U_2 with base U_1, associated subgroups $\mathrm{gp}(y_i)$, $\mathrm{gp}(y_i^2)$ $(i = 0, \ldots, m)$, and stable letters $t_0, \ldots, t_m$ as follows:

$$U_2 = \langle\, U_1, t_0, \ldots, t_m \; ; \; t_0 y_0 t_0^{-1} = y_0^2, \ldots, t_m y_m t_m^{-1} = y_m^2 \,\rangle\,.$$

By Britton's Lemma $H = \mathrm{gp}(t_0, \ldots, t_m)$ is free on $t_0, \ldots, t_m$ and hence

$$t_i \longmapsto t_i^2 \qquad (i = 0, \ldots, m)$$

defines an isomorphism from H onto $K = \mathrm{gp}(t_0^2, \ldots, t_m^2)$. So we can form another HNN extension U_3:

$$U_3 = \langle\, U_2, z \; ; \; z t_i z^{-1} = t_i^2 \ (i = 0, \ldots, m) \,\rangle\,.$$

Next let V_1 be the free group on r. Form the HNN extension

$$V_2 = \langle V_1, s \, ; \, srs^{-1} = r^2 \rangle \, .$$

Again s is of infinite order. So we can form another HNN extension V_3:

$$V_3 = \langle V_2, t \, ; \, tst^{-1} = s^2 \rangle \, .$$

Now a major move. Put

$$W = \{ \, U_3 * V_3 \, ; \, r = z \, , \, t = [w, y_0] \, \} \, .$$

Observe that if $w \neq 1$, then it follows from Britton's Lemma applied to U_3, that

$$\text{gp}(z, [w, y_0]) \quad \text{is free on} \quad z \text{ and } [w, y_0].$$

Similarly, it follows from another application of Britton's Lemma, this time applied to V_3, that

$$\text{gp}(r, t) \quad \text{is free on} \quad r, \, t.$$

This means that, in the event that $w \neq 1$, W is an amalgamated product, where the amalgamated subgroup is a free group of rank two and therefore contains G_2. It follows that we have proved that

$$\text{if } w \neq 1, \text{ then } W \text{ does not have } \mathcal{M}.$$

Let's see what happens if $w = 1$. Tracing our way back through the construction we find that

$$[w, y_0] = 1 \implies t = 1 \implies s = 1 \implies r = 1 \implies z = 1 \implies t_0 = \ldots = t_m = 1$$
$$\implies y_0 = \ldots = y_m = 1.$$

In other words, if $w = 1$ then $W = \{1\}$.

Now put

$$G_w = W * G_1.$$

It follows that G_w has $\mathcal{M}$ if and only if $w = 1$. Since there is no algorithm which decides whether or not $w = 1$, there is no algorithm which decides whether or not G_w has $\mathcal{M}$. This completes the proof of the Adian-Rabin theorem. ∎

It is worth noting that, in the case where $\mathcal{M}$ is the property of being of order 1, $G_1 = 1$ and so, in this instance, we have concocted a family of finitely presented groups G_w, such that $G_w = 1$, if and only if $w = 1$, where again w ranges over the words in the generators of a fixed finitely presented group with an unsolvable word problem. This class of groups can be used to obtain further negative algorithmic results about finitely presented groups, as we will see from Theorem 6, below.

The following exercise will be of use in the proof of the next theorem.

Exercise 2 Let $G = A * B$ $(A \neq 1 \neq B)$. Prove the following:
(i) there exists an element $g \in G$ with infinite cyclic centralizer;
(ii) $\zeta G = 1$;
(iii) G is directly indecomposable (use (i)).

The following theorem is a simple application of the existence of the groups G_w, described just before Exercise 2.

Theorem 6 There is no algorithm which decides whether or not any finitely presented group
(i) is isomorphic to its direct square;
(ii) has an infinite automorphism group;
(iii) is a non-trivial free product;
(iv) is centerless;
(v) has an infinitely generated subgroup.

To prove (i) for instance, note that, by part (iii) of Exercise 2, $G_w * G_w$ is isomorphic to a direct square of itself, if and only if $G_w = 1$. The other parts of the proof are left as an exercise for the reader.

Here is one further illustration of this technique. In order to explain let me remind you of some definitions. Suppose G is any group. Then (see the book by P. J. Hilton and U. Stammbach: *A course in Homological Algebra*, **Graduate Texts in Mathematics 4**, Springer-Verlag, New York-Heidelberg-Berlin (1971)) define

$$H_1(G, \mathbf{Z}) = G_{ab} .$$

$H_1(G, \mathbf{Z})$ is the first homology group of G with coefficients in the additive group $\mathbf{Z}$ of integers. It is only one of a whole sequence of groups, starting with

$$H_0(G, \mathbf{Z}) = \mathbf{Z} , \ H_1(G, \mathbf{Z}) , \ H_2(G, \mathbf{Z}) , \ \dots .$$

The homology group which seems to be of most direct interest in the study of finitely presented groups, is $H_2(G, \mathbf{Z})$, which, like $H_1(G, \mathbf{Z})$, has a group-theoretic description. Indeed, if we express G in the form

$$G \cong F/K ,$$

where F is a free group, then

$$H_2(G, \mathbf{Z}) = (F' \cap K)/[F, K] .$$

This description of $H_2(G, \mathbf{Z})$ seems to depend on the representation of G as F/K. However this is not the case, as the following exercise shows.

Exercise 3 *(i) Use Tietze transformations to prove that if*

$$G \cong F/K \cong E/S$$

where E and F are free, then

$$(F' \cap K)/[F, K] \cong (E' \cap S)/[E, S] .$$

(ii) Prove that $H_2(G, \mathbf{Z}) = 0$ *if G is free.*

We have the following simple

Lemma 4 *Suppose that G is finitely presented. Then* $H_2(G, \mathbf{Z})$ *is finitely generated. Indeed if*

$$G = \langle\, x_1, \ldots, x_m \,;\, r_1, \ldots, r_n \,\rangle \qquad (m, n < \infty),$$

then $H_2(G, \mathbf{Z})$ *is an abelian group that can be generated by n elements and hence is finitely generated.*

Proof We express G in the form

$$G \cong F/K$$

where

$$F = \langle\, x_1, \ldots, x_m \,\rangle$$

and

$$R = \mathrm{gp}_F(r_1, \ldots, r_n) .$$

Then

$$R/[F, K] = \mathrm{gp}(\, r_1[F, K], \ldots, r_n[F, K] \,) .$$

Hence $(F' \cap K)/[F, K]$ can be generated by n elements, by the basis theorem for finitely generated abelian groups. ∎

Thus to each finitely presented group G we can associate two finitely generated abelian groups, $H_1(G, \mathbf{Z})$ and $H_2(G, \mathbf{Z})$. The first of these is computable, by the basis theorem for finitely generated abelian groups. Somewhat surprisingly, the second of these is not. This result is due to Cameron Gordon (see the reference to C.F. Miller III in Chapter I). In order to prove this assertion, we again make use of the groups G_w cited just before Exercise 2. Notice that each of the G_w involved, can be presented on the same number of generators and the same number of relations:

$$G_w = \langle\, x_1, \ldots, x_m \,;\, r_1, \ldots, r_n \,\rangle .$$

Of course these presentations depend on w. Now form the free product of each G_w with the free group on s and t:

$$E_w = G_w * \langle s, t \rangle .$$

Next observe that if $w \neq 1$, then

$$\mathrm{H}_w = \mathrm{gp}\big(\, [w, t] \, , \, s[w, t]s^{-1} \, , \, \dots \, , \, s^\alpha [w, t]s^{-\alpha} \, \big)$$

is free of rank $\alpha + 1$, where we here choose

$$\alpha = 2m + 4 .$$

Let $\overline{E}_w$ be an isomorphic copy of E_w and let $\overline{\mathrm{H}}_w$ be the corresponding copy of H_w in $\overline{E}_w$. Consider next the amalgamated product

$$P_w = \big\{ \, E_w * \overline{E}_w \, ; \, \mathrm{H}_w = \overline{\mathrm{H}}_w \, \big\} .$$

There is a sequence, called the Mayer-Vietoris sequence, which describes the homology groups of P_w by a long exact sequence, part of which looks like this:

$$\dots \longrightarrow \mathrm{H}_2(P_w, \mathbf{Z}) \longrightarrow \mathrm{H}_1(\mathrm{H}_w, \mathbf{Z}) \overset{\gamma}{\longrightarrow} \mathrm{H}_1(E_w, \mathbf{Z}) \oplus \mathrm{H}_1(\overline{E}_w, \mathbf{Z}) \longrightarrow \dots .$$

Now $\mathrm{H}_1(\mathrm{H}_w, \mathbf{Z})$ is free abelian of rank $2m + 5$ and $\mathrm{H}_1(E_w, \mathbf{Z}) \oplus \mathrm{H}_1(\overline{E}_w, \mathbf{Z})$ can be generated by $2m + 4$ elements. Hence $\ker \gamma \neq 0$. This means that $\mathrm{H}_2(P_w, \mathbf{Z}) \neq 0$. This is all predicated on the assumption that $w \neq 1$. If $w = 1$, then $G_w = 1$ and P_w is free. Hence $\mathrm{H}_2(P_w, \mathbf{Z}) = 0$ by part (ii) of Exercise 3. Thus we have proved:

Theorem 7 *There is no algorithm whereby one can decide whether or not any finitely presented group G has $\mathrm{H}_2(G, \mathbf{Z}) = 0$.*

Here is one last application of HNN extensions, providing us with perhaps the simplest examples of non-hopfian groups.

Theorem 8 *The group*

$$G = \langle \, a, t \, ; \, t^{-1}a^2 t = a^3 \, \rangle$$

is non-hopfian.

Proof G is of course an *HNN* extension with a single stable letter t and an infinite cyclic base $\langle a \rangle$. Consider the map

$$\varphi \, : \, a \longmapsto a^2 \, , \, t \longmapsto t .$$

Since
$$t^{-1}(a^2)^2 t = (a^2)^3$$

it follows from Dyck's Theorem that φ defines a homomorphism, again denoted φ, of G into G. Observe that
$$G = \operatorname{gp}(a^2, t) \, .$$

So φ is onto. Now consider

$$g = (a^{-1}t^{-1}at)^2 a^{-1} = a^{-1}t^{-1}ata^{-1}t^{-1}ata^{-1} \, .$$

There is no pinch in g. So, by Britton's Lemma, $g \neq 1$. But

$$g\varphi = (a^{-2}t^{-1}a^2t)^2 a^{-2} = a^2 a^{-2} = 1 \, .$$

So $G \cong G/\ker \varphi$ is non-hopfian. ∎

This theorem is due to G. Baumslag and D. Solitar (see the references cited in the book by Lyndon and Schupp).

We now construct some finitely generated groups which are not finitely presented, by means of generalised free products.

Theorem 9 *Let A and B be finitely presented groups. Then*
$$G = A \underset{H}{*} B$$
is finitely presented if and only if H is finitely generated.

Proof One part of the theorem is obvious. To prove that G is not finitely presented if H is not finitely generated, let

$$A = \langle X; R \rangle \quad , \quad B = \langle Y; S \rangle$$

be finite presentations for A and B. Suppose that

$$H = \operatorname{gp}(h_1(\underline{x}), h_2(\underline{x}), \ldots)$$

where $h_i(\underline{x})$ denotes, as usual a word in the generators of X of A. There exist corresponding words $k_i(\underline{y})$ in the generators Y of B such that

$$h_i(\underline{x}) = k_i(\underline{y}) \qquad (i = 1, 2, \ldots) \, .$$

So G can be presented in the form

$$G = \langle X \cup Y; R \cup S \cup \{ h_i(\underline{x}) k_i(\underline{y})^{-1} \mid i = 1, 2, \ldots \} \rangle \, .$$

Since $X \cup Y$ is finite, it follows from Neumann's theorem, that if G is finitely presented, then it can be presented in the form

$$G = \langle\, X \cup Y \,;\, R \cup S \cup \{\, h_i(\underline{x})k_i(\underline{y})^{-1} \mid i = 1, \ldots, n \,\} \,\rangle \qquad (4)$$

for some choice of the positive integer n. Since H is not finitely generated, it follows that

$$H_1 = \mathrm{gp}\big(\, h_1(\underline{x}), \ldots, h_n(\underline{x}) \,\big) \neq H \;.$$

Therefore, there exists an element $h(\underline{x}) \in H$, with $h(\underline{x}) \notin H_1$. Thus if $k(\underline{y})$ is the element in B corresponding to $h(\underline{x})$, under the given isomorphism between H and K, then

$$k(\underline{y}) \notin K_1 = \mathrm{gp}\big(\, k_1(\underline{y}), \ldots, k_n(\underline{y}) \,\big) \;.$$

Let us think of (4) as presenting a group, supposedly G, but more safely denoted by $\widetilde{G}$:

$$\widetilde{G} = \langle\, X \cup Y \,;\, R \cup S \cup \{\, h_i(\underline{x})k_i(\underline{y})^{-1} \mid i = 1, \ldots, n \,\} \,\rangle \;.$$

Now $\widetilde{G}$ is, by its very presentation, an amalgamated product:

$$\widetilde{G} = \{\, A * B \,;\, H_1 = K_1 \,\} \;.$$

Consequently

$$h(\underline{x})k(\underline{y})^{-1} \neq 1$$

in $\widetilde{G}$ because it is a strictly alternating product. On the other hand we have

$$h(\underline{x}) = k(\underline{y})$$

in G. So $\widetilde{G}$ is not isormorphic to G and therefore G is not finitely presented, as claimed. ∎

Exercises 4

(i) Suppose that $E = \langle\, B, t \,;\, tHt^{-1} = K \,\rangle$ is an HNN extension with finitely presented base B. Prove that E is finitely presented if and only if H is finitely generated.

(ii) If $F = \langle a, b \rangle$, then $G = F \underset{F'}{*} F$ is not finitely presented.

(iii) $A = \langle\, a, s \,;\, a^s = a^2 \,\rangle$, $B = \langle\, b, t \,;\, b^t = b^2 \,\rangle$, then $G = \{\, A * B \,;\, \mathrm{gp}_A(a) = \mathrm{gp}_B(b) \,\}$ is not finitely presented.

As one last illustration, here is an example of a finitely presented group with a finitely generated subgroup which is not finitely presented.

Example 1 *Let $F_i = \langle a_i, b_i \rangle$ $(i = 1, 2)$ be free of rank two. Consider*

$$D = F_1 \times F_2$$

the direct product of F_1 and F_2. Then D is clearly finitely presented. (In the discussion that follows we denote the elements of D by pairs (u, v), where $u \in F_1$ and $v \in F_2$.) We claim that

$$H = \mathrm{gp}\big(\alpha = (a_1, a_2),\, \beta = (b_1, 1),\, \gamma = (1, b_2)\big)$$

is not finitely presented. In order to prove this assertion, first observe (and also prove) that H can be presented as follows:

$$H = \langle \alpha, \beta, \gamma;\, [\beta^{\alpha^i}, \gamma^{\alpha^j}] = 1\ (i, j \in \mathbf{Z}) \rangle .$$

Suppose H were finitely presented. Then, by Neumann's theorem, it has a presentation of the form

$$H = \langle \alpha, \beta, \gamma;\, [\beta^{\alpha^i}, \gamma^{\alpha^j}] = 1\ (-N \le i, j \le N) \rangle$$

for some positive integer N. We compute now a presentation for

$$K = \mathrm{gp}_H(\beta, \gamma) .$$

Notice that $H/K = \langle \alpha K \rangle$ is infinite cyclic. So we have a ready-made right Schreier transversal $S = \{\alpha^i \mid i \in \mathbf{Z}\}$ for K in H. Put

$$\beta_i = \alpha^i \beta \alpha^{-i},\quad \gamma_i = \alpha^i \gamma \alpha^{-i} \qquad (i \in \mathbf{Z}).$$

Then it follows that K has the following presentation:

$$K = \langle \ldots, \beta_i, \ldots, \ldots, \gamma_j, \ldots;\, [\beta_i, \gamma_j] = 1(i, j \in \mathbf{Z},\, |i - j| \le 2N) \rangle.$$

Put

$$X = \mathrm{gp}(\beta_{-N}, \ldots, \beta_N),\ Y = \mathrm{gp}(\gamma_{-N}, \ldots, \gamma_N) .$$

Then X and Y are free on the exhibited generators and

$$A = \mathrm{gp}(X, Y) = X \times Y .$$

Similarly

$$B = \mathrm{gp}(\alpha X \alpha^{-1}, \alpha Y \alpha^{-1}) = \alpha X \alpha^{-1} \times \alpha Y \alpha^{-1} .$$

Observe that $\mathrm{gp}(A, B)$ is an amalgamated product:

$$\mathrm{gp}(A, B) = \{A * B; \beta_{-N+1} = \beta_{-N+1}, \ldots, \beta_N = \beta_N, \gamma_{-N+1} = \gamma_{-N+1}, \ldots, \gamma_N = \gamma_N\}.$$

But then $[\beta_{-N}, \gamma_{N+1}]$ is not equal to 1 in K and hence not equal to 1 in H. Since this is patently false, it follows that H is not finitely presented, as claimed.

4   Some word, conjugacy and isomorphism problems

Definition 2 *Let A be a group given by a finite presentation and let H be a finitely generated subgroup of A given by a finite set of generators, each of which comes expressed as a word in the given generators of A. Then we say that the occurence problem, or extended word problem, for A relative to H, is solvable, if there exists an algorithm such that for each word w in the generators of A, we can decide whether or not $w \in H$. In this case, it follows that there is an effective method to exhibit w as a word in the generators of H.*

Proposition 1 *Suppose*

$$G = \{ A * B ; H \stackrel{\varphi}{=} K \}$$

is an amalgamated product, where A and B are given by finite presentations. In addition, suppose that H and K are finitely generated subgroups of A and B, that they come equipped with equally indexed finite sets of generators, and that φ is defined by sending each generator of H to the correspondingly indexed generator of K. Furthermore, suppose that the extended word problem for A relative to H is solvable and that the extended word problem for B relative to K is solvable. If the word problem for H is solvable, using the given generators, then so is the word problem for G.

It is worth pointing out that the subgroup H need not be finitely presented. The formulation of the proposition is a great deal more difficult than the proof, which follows immediately from the fact that a strictly alternating product in G is different from 1.

There are two classes of groups for which the extended word problem is solvable. In order to explain, we need another definition.

Definition 3 *Let A be a group and let H be a subgroup of A. We term H finitely separable in A if, for each $a \in A$, $a \notin H$, there exists a normal subgroup N of A of finite index such that*

$$a \notin NH .$$

Lemma 5 *Let A be a finitely presented group, H a finitely generated subgroup of A which is finitely separable in A. Then the extended word problem for A relative to H is solvable.*

Proof We begin by listing the finite quotients of A and the elements of H. Then, for any given $a \in A$, we will either find that the image of a is not in the image H in some finite quotient of A, or else we will find that $a \in H$.   ∎

Lemma 6 (Toh) *Let A be a finitely generated nilpotent group, H a finitely generated subgroup of A. Then H is finitely separable in A.*

The proof follows easily by induction on the class of A and is left to the reader.

Lemma 7 *Let A be a finitely generated free group, H a finitely generated subgroup of A. Then H is finitely separable in A.*

Proof By Marshall Hall's theorem, there exists a subgroup J of A of finite index such that H is a free factor of J. Let $a \in A$, $a \notin J$. Since J is of finite index in A, there exists a normal subgroup N of A contained in J of finite index in A. Then $a \notin NH$. So we are half-way to proving that H is finitely separable in A.

Suppose next that $a \in J$, $a \notin H$. Now

$$J = H * L .$$

Now H and L are residually finite. So we can choose normal subgroups H_1 of H and L_1 of L in such a way that H/H_1 and L/L_1 are finite, and so that the canonical image $\bar{a}$ in

$$\bar{J} = H/H_1 * L/L_1$$

is not in $\bar{H} = H/H_1$. Notice that $\bar{J}$ is the free product of the two finite groups $\bar{H}$ and $\bar{L} = L/L_1$. Using the Reidemeister-Schreier method it is not hard to see that the kernel K of the homomorphism of $\bar{J}$ onto $\bar{H} \times \bar{L}$ is free. Thus $\bar{J}$ is a finite extension of a free group and hence residually finite. But $\bar{H}$ is a finite subgroup of $\bar{J}$. Therefore we can find a normal subgroup $\bar{S}$ of finite index in $\bar{J}$ such that

$$\bar{w} \notin \bar{S}\bar{H} .$$

Pulling this information back into J yields a normal subgroup S of finite index in J such that

$$a \notin SH.$$

Now S is of finite index in J, hence of finite index in A. Therefore the conjugates of S intersect in a subgroup T of S which is of finite index in A and normal in A. But notice that $a \notin TH$ since $TH \leq SH$. ∎

On combining Proposition 1 and Lemma 7, it follows that we have proved the

Theorem 10 *The free product of two finitely generated free groups (or two finitely generated nilpotent groups) with a finitely generated subgroup amalgamated has a solvable word problem.*

Corollary 1 (Dehn) *The fundamental groups of closed two-dimensional surfaces have solvable word problem.*

Chapter VII
Groups acting on trees

1 Basic definitions

The exposition in this chapter is based on, and sometimes follows very closely, the book by Jean-Pierre Serre: *Trees*, **Translated from the French by John Stillwell**, Springer-Verlag, Berlin, Heidelberg, New York (1980). The reader should consult this work for more details, if needed.

Definition 1 *A graph X is a pair of sets, $V = V(X) \neq \emptyset$ and $E = E(X)$, termed the vertices and edges of X, equipped with three maps*

$$o \, : \, E \longrightarrow V \, , \qquad t \, : \, E \longrightarrow V \, , \qquad ^{-} \, : \, E \longrightarrow E$$

satisfying the following conditions: if $e \in E$, then
(i) $e \neq \bar{e}$ and $\bar{\bar{e}} = e$ (i.e., the map $^{-}$ is of order two and is fixed point free);
(ii) $o(e) = t(\bar{e})$.

We term $t(e)$ the *terminus* of e, $o(e)$ its *origin* and $\bar{e}$ the *inverse* of e. Sometimes we refer to $o(e)$ and $t(e)$ as the *extremities* of e. It is possible for $o(e) = t(e)$ and in this case e is termed a *loop*. Two distinct vertices are termed *adjacent* if they are the extremities of some edge.

Graphs are often represented by diagrams in the plane, the vertices by points and the edges by line segments joining its extremities. We usually affix to such line segments an arrow, whose direction emanates from the origin of the edge and is directed towards its terminus. We customarily omit one of e, $\bar{e}$. Usually diagrams are drawn in such a way that the graphs can be reconstructed from them without ambiguity.

We take for granted the usual notions involving *morphisms of graphs*, with the *automorphism group* Aut X of the graph X consisting of the invertible morphisms $X \longrightarrow X$, using composition as the binary operation.

Definition 2 *A group G acts on a graph X if it comes equipped with a homomorphism*

$$\varphi \, : \, G \longrightarrow \text{Aut } X \, .$$

We often denote the image of $v \in V(X)$ under the action of $g \in G$ by gv.

Examples 1 *(1) Let G be a group, S a set of generators of G. We define the Cayley graph $X = X(G, S)$ of G relative to S by*
(i) $V(X) = G$;
(ii) $E(X)$ is the disjoint union of the sets $G \times S$ and $S \times G$;
(iii) $o(g, s) = g$; $t(g, s) = gs$; $\overline{(g, s)} = (s, g)$, $\overline{(s, g)} = (g, s)$.

Notice that G acts on $X(G, S)$ by left multiplication:

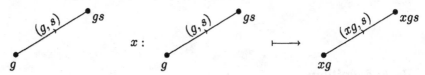

(2) $G = \{1\}$. Then $X(G, \{1\})$ is a loop: 1 $\bigcirc$ (1,1)

Notice that despite the notation, we have the inverse edge $\overline{(1, 1)}$ which is different from $(1, 1)$.

(3) $G = \langle a; a^n = 1 \rangle$ $(n \in \{1, 2, \ldots\})$, $S = \{a\}$. Then $X(G, S)$ is what is often referred to as a circuit of length n:

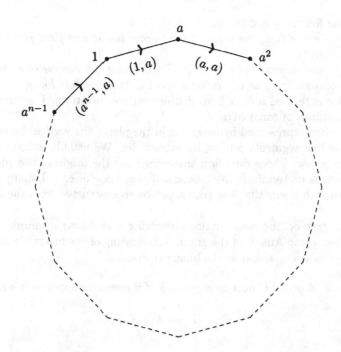

(4) $G = \langle a \rangle$, $S = \{a\}$. Then $X(G, S)$ can be drawn as follows:

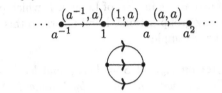

(5)

(6) *A point:*

(7) *A segment:*

(8) *A path of length n: P_n*

Definition 3 *Let X be a graph. A morphism $f : P_n \longrightarrow X$ is again termed a path (of length n). We term $f(0)$ and $f(n)$ the extremities of f in X and say f goes from $f(0)$ to $f(n)$. The path f is called a closed path if $f(0) = f(n)$.*

Definition 4 *A graph is said to be connected if any two vertices are the extremities of at least one path. The maximal connected subgraphs (under inclusion) are called the connected components of a given graph.*

A path $f : P_n \longrightarrow X$ can be identified as a succession of edges

$$e_1, \ldots, e_n \quad \text{where} \quad e_i = f\left(\underset{i-1}{\bullet} \longrightarrow \underset{i}{\bullet} \right) \quad (i = 1, \ldots, n).$$

A consecutive pair of edges e_i, e_{i+1} is termed a *backtracking* if $e_{i+1} = \overline{e_i}$. A path f with extremities P and Q is termed a *geodesic* if it is of minimal length, i.e., any path with the same extremities has at least as great a length.

Definition 5 *A graph X is termed a tree if it is connected and every closed path in X of positive length contains a backtracking.*

Examples of trees

(1)

(2) $\cdots \bullet \longrightarrow \bullet \longrightarrow \bullet \longrightarrow \bullet \longrightarrow \bullet \cdots$ *an infinite path.*

A rich source of trees comes from graphs of free groups. Here is a simple lemma which clarifies the kinds of Cayley graphs one can get.

Lemma 1 *Let G be a group, S a set of generators of G, $X = X(G, S)$ the Cayley graph of G relative to S. Then the following hold:*
(i) X is connected;
(ii) X contains a loop if and only if $1 \in S$;
(iii) G acts on X without inversion, i.e., $ge \neq \overline{e}$ ($e \in E(X)$, $g \in G$).

The proof of Lemma 1 is straightforward. The one feature I want to emphasize is that of an *inversion*, defined above in *(iii)* of Lemma 1, which plays a crucial role in constructing quotients by group actions. I will return to this in a moment. But first let me look at another example.

Example 2 *Let G be free on x,y, let $S = \{x,y\}$ and let $X = X(G,S)$. Then we claim X is a tree. To begin with, of course, by Lemma 1, X is connected and contains no loops and G acts on X without inversion. Here is an attempt to draw X in the plane:*

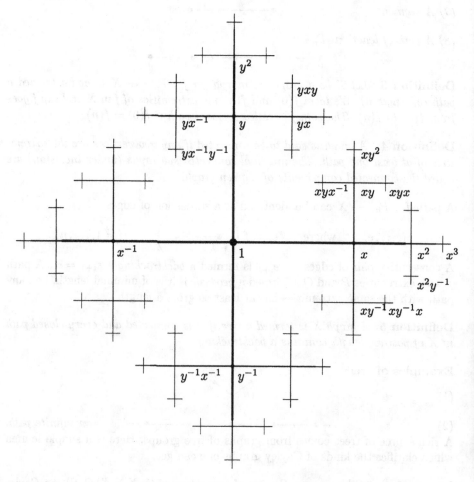

It is not hard to see that X is a tree. For suppose we have a closed path in X of length $n > 0$, beginning and ending at w. Then it can be viewed as a succession of edges

$$e_1, \ldots, e_n$$

where

$$o(e_1) = w , \quad t(e_n) = w .$$

Now notice that

$$t(e_1) = w z_1 = o(e_2) , \quad t(e_2) = w z_1 z_2 = o(e_3) , \quad \dots , \quad t(e_n) = w z_1 \dots z_n .$$

Here each $z_i \in \{x, y, x^{-1}, y^{-1}\}$. *Since* $t(e_n) = w$,

$$z_1 \dots z_n = 1 .$$

So $z_1 \dots z_n$ *is not a reduced word, i.e., for some* i,

$$z_i z_{i+1} = 1 .$$

This means that the given closed path in X *contains a backtracking, as required.*

The relevance of groups acting without inversion on a graph is clarified next by Lemma 2. But first we need some additional notions.

Definition 6 *An orientation of a graph* X *is a decomposition*

$$E = E_+ \cup E_-$$

of $E = E(X)$ *into two disjoint sets* E_+ *and* E_- *such that*

$$\overline{E_+} = E_- , \quad \overline{E_-} = E_+ .$$

Every graph has such an orientation since the map $\overline{}$ is of order two and is fixed point free. Thus every orbit has two elements and we can take for E_+ any set of representatives of these orbits and for E_- the complementary set. Sometimes we then refer to the edges in E_+ as *positive edges*, those in E_- as negative edges. A graph X with a prescribed orientation is usually referred to as an *oriented* graph. Most of our diagrams represent oriented graphs.

Definition 7 *A morphism of graphs* $f : X \longrightarrow X'$ *is termed orientation preserving if there exists orientations of* X *and* X' *which are preserved by* f, *i.e., if there exists an orientation*

$$E(X) = E_+(X) \cup E_-(X)$$

of X *and an orientation*

$$E(X') = E_+(X') \cup E_-(X')$$

of X' *such that*

$$f\big(E_+(X)\big) \subseteq E_+(X'), \quad f\big(E_-(X)\big) \subseteq E_-(X') .$$

Lemma 2 *Suppose that* G *acts on a graph* X. *Then* G *acts without inversion if and only if* G *is orientation preserving, i.e., if there exists an orientation* $E = E_+ \cup E_-$ *of* X *preserved by* G.

Proof Decompose E into disjoint G-orbits:

$$E = \bigcup \langle e \rangle$$

where

$$\langle e \rangle = Ge \quad \text{and} \quad \langle v \rangle = Gv \quad (v \in V(X)) .$$

If G acts without inversion, $^-$ acts as an involution on the orbits $\langle e \rangle$ and we can therefore partition E into $E = E_+ \cup E_-$ in such a way that

$$E_+ = \bigcup \langle e \rangle \quad , \quad E_- = \bigcup \langle f \rangle .$$

The converse is an immediate consequence of this definition. ∎

Definition 8 *Let G act without inversion on X. Define the quotient graph $G \backslash X$ as follows:*

$$o\langle e \rangle = \langle o(e) \rangle \quad , \quad t\langle e \rangle = \langle t(e) \rangle \quad , \quad \overline{\langle e \rangle} = \langle \bar{e} \rangle .$$

The point here is that $^-$ makes sense since G acts without inversion on X.

The map

$$f \; : \; X \longrightarrow G \backslash X$$

defined by

$$f \; : \; e \longmapsto \langle e \rangle \quad , \quad v \longmapsto \langle v \rangle$$

is a morphism of graphs.

Example 3 *Let G be free of rank two on x and y, $S = \{x, y, x^{-1}, y^{-1}\}$ and $X = X(G, S)$ as before. So X is a tree. We now compute $G \backslash X$.*

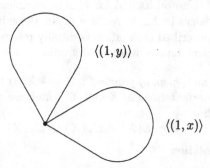

So $G \backslash X$ is a two-leaved rose, reflecting the fact that G is free of rank two. We shall obtain a general stucture theorem for groups acting on a tree. The clue to the structure of these groups will come from an examination of the corresponding quotient graphs, from which we will be able to reconstruct G itself. In order to try to motivate what follows, let me digress for a few minutes and take you on a quick trip through covering space theory.

2 Covering space theory

In this quick trip though covering space theory, I will take for granted many elementary notions of topology, such as a *topological space*, *arc-wise connectedness* and the *fundamental group* $\pi_1(X, *)$ of a space X based at a point $*$. All spaces will be *Hausdorff*, i.e., distinct points have disjoint neighbourhoods, and *locally arc-wise connected*, i.e., if V is an open set containing a point x, there exists an open subset U contained in V and containing x such that any pair of points in U have a path in U joining them.

Definition 9 *Let X and $\widetilde{X}$ be two arc-wise connected, locally arc-wise connected spaces, $p : \widetilde{X} \longrightarrow X$ a continuous map. We term $\left(\widetilde{X}, p \right)$ a covering space of X if*
(i) p is onto;
(ii) each $x \in X$ has an open neighborhood U such that $p^{-1}(U)$ is a disjoint union of open sets homeomorphic via p to U. (Such sets U are usually called elementary neighbourhoods.)

Examples 4

(1) $X = S^1$
$= \{ z \in \mathbf{C} \mid |z| = 1 \}$, $\widetilde{X} = \mathbf{R}^1$, $p : \widetilde{X} \longrightarrow X$, $r \longmapsto e^{2\pi i r}$.

(2) $X = \mathbf{P}_{\mathbf{R}}^2$ the real projective plane. Points of X are lines in $\mathbf{R}^3$ through 0. Open sets in X are open "cones" of lines:

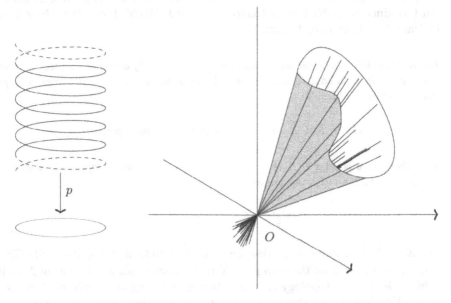

$$\widetilde{X} = S^2 = \{ (x, y, z) \in \mathbf{R}^3 \mid x^2 + y^2 + z^2 = 1 \},$$
$$X = S^2 = \{ (x, y, z) \in \mathbf{R}^3 \mid x^2 + y^2 + z^2 = 1 \},$$

$p : \widetilde{X} \longrightarrow X$ *maps a point* $(x, y, z) \in \widetilde{X}$ *to the line through 0 passing through* (x, y, z). *Notice that* $p^{-1}(line) = $ *two points, i.e., p is what is usually referred to as a 2-sheeted covering.*

(3) $X = T^2 = S^1 \times S^1$, *a torus,*
$\widetilde{X} = \mathbf{R}^2$
$p : \widetilde{X} \longrightarrow X$
$(r_1, r_2) \longmapsto \left(e^{2\pi i r_1}, e^{2\pi i r_2}\right)$

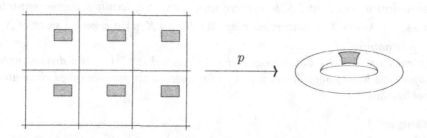

A covering space of a space X is termed a *universal covering space* if it is *simply connected*, i.e., if its fundamental group is of order 1. Such universal covering spaces are, in a sense, unique and so are usually referred to as *the* universal covering of the space X. (See the book by William S. Massey: **Algebraic Topology: An Introduction**, published by Harcourt, Brace & World, Inc. (1968), New York, Chicago, San Francisco, Atlanta.)

Definition 10 *Let $\widetilde{X}$ be an arc-wise connected, locally arc-wise connected space. Then a group G is said to act properly discontinuously on $\widetilde{X}$ if it comes equipped with a homomorphism*

$$\varphi : G \longrightarrow \text{Aut } \widetilde{X} \quad (= \text{the group of homeomorphisms of } \widetilde{X})$$

such that every point $\widetilde{x} \in \widetilde{X}$ is contained in a so-called proper open neighbourhood V, i.e., V has the following property

$$V \cap gV = \emptyset \qquad (g \in G, g \neq 1) .$$

Let $X = G \backslash \widetilde{X}$ denote the quotient space of $\widetilde{X}$. Points of X are the orbits $G\widetilde{x}$ of points $\widetilde{x} \in \widetilde{X}$. Let p be the map from $\widetilde{X}$ to X which sends an element of $\widetilde{X}$ to its orbit under G. The topology of X is obtained by taking as a basis for X the sets $U = p(V)$ where V is a proper open set in $\widetilde{X}$. Then the map $p : \widetilde{X} \longrightarrow X$ makes $\left(\widetilde{X}, p\right)$ a covering space of X.

Theorem 1   *Let $\widetilde{X}$ be simply connected, arc-wise connected and locally arc-wise connected. Suppose that G acts properly discontinuously on $\widetilde{X}$. Then*

$$G \cong \pi_1\left(G\backslash\widetilde{X}, *\right) \ .$$

In other words, if G acts properly discontinuously on a space $\widetilde{X}$ with the right properties, then we can recapture G from the fundamental group of the quotient space $G\backslash\widetilde{X}$ of $\widetilde{X}$. This motivates, roughly speaking, what we will do here. More precisely, suppose that a group G acts without inversion on a tree X. We form the quotient graph $G\backslash\widetilde{X}$, keeping track of the stabilisers of some of the vertices and edges in X under the action of G. This information is codified in terms of a so-called *graph of groups*, a subject to which we will turn next. The group G is then recaptured using this information.

3   Graphs of groups

The following definition turns out to be an important tool in studying groups acting without inversion on a tree.

Definition 11   *A pair $(\mathcal{G}, Y)$, satisfying the following conditions, is termed a graph of groups:*
(1) Y is a connected graph;
(2) $\mathcal{G}$ is a mapping from $V(Y) \cup E(Y)$ into the class of all groups;
(3) the image of $P \in V(Y)$ under $\mathcal{G}$ is denoted by G_P and is termed the vertex group at P or simply a vertex group;
(4) the image of $y \in E(Y)$ under $\mathcal{G}$ is denoted by G_y and is termed the edge group at y or simply an edge group;
(5) $G_y = G_{\overline{y}}$ for every $y \in E(Y)$;
(6) each edge group G_y comes equipped with a monomorphism

$$G_y \longrightarrow G_{t(y)} \qquad \text{denoted by} \qquad a \longmapsto a^y \quad (a \in G_y) \ .$$

We can now amplify a little the comment made above. To this end, suppose that G acts without inversion on a tree X. We then associate with this action a graph $(\mathcal{G}, Y)$ of groups. The graph Y is the quotient graph $G\backslash X$. The vertex groups and edge groups of the graph $(\mathcal{G}, Y)$ of groups, are stabilisers of a carefully selected set of edges and vertices of X. Now given any graph $(\mathcal{G}, Y)$ of groups we associate to it a group which is analogous to the fundamental group of a topological space, termed its *fundamental group* and denoted by $\pi_1(\mathcal{G}, Y)$. This group $\pi_1(\mathcal{G}, Y)$ is constructed from the vertex groups and edge groups of $(\mathcal{G}, Y)$, by using amalgamated products and HNN extensions. The point here is that if we go back to our given group G, acting without inversion on a tree X, and construct the corresponding graph of groups $(\mathcal{G}, Y)$, then it turns out that

$$G \cong \pi_1(\mathcal{G}, Y).$$

This yields the desired structure theorem for groups acting without inversion on a tree. The theory of groups acting on trees is completed by proving that the fundamental group $\pi_1(\mathcal{G}, Y)$ of a graph of groups $(\mathcal{G}, Y)$ acts on a tree. Consequently, free groups, amalgamated products, and HNN extensions act on trees. Notice that if a group acts without inversion on a tree, so does every one of its subgroups. This means, e.g., that a subgroup of an amalgamated product or of an HNN extension is the fundamental group of a graph of groups. It follows that this theory yields subgroup theorems for amalgamated products and HNN extensions. These subgroup theorems were first obtained by A. Karrass and D. Solitar: *The subgroups of a free product of two groups with an amalgamated subgroup*, **Transactions of the American Math. Soc. vol. 150,** pp. 227-250 (1970).

We give now some important examples of graphs of groups and describe, without any justification, the associated fundamental groups.

Examples 5
(1) A loop of groups.

$$Y \ : \qquad P \bigcirc y \qquad\qquad (\mathcal{G}, Y) \ : \qquad G_P \bigcirc G_y$$

So a loop of groups consists of a group G_P, a second group G_y and two monomorphisms of G_y into G_P:

$$G_y \longrightarrow G_P \ , \qquad G_y \longrightarrow G_P$$
$$a \longmapsto a^y \qquad\qquad a \longmapsto a^{\bar{y}} .$$

Here the fundamental group $\pi_1(\mathcal{G}, Y)$ turns out to be an HNN extension with one stable letter, base group G_P and associated subgroups the two images of G_y.

(2) A segment of groups.

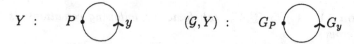

$$Y \ : \quad \underset{P \quad\ Q}{\bullet \overset{y}{\longrightarrow} \bullet} \ ; \ (\mathcal{G}, Y) \quad \underset{G_P \quad G_Q}{\overset{G_y}{\bullet \longrightarrow \bullet}} \qquad\qquad G_y \begin{array}{c} \nearrow a \longmapsto a^y\ \ G_Q \\ \searrow a \longmapsto a^{\bar{y}}\ G_P \end{array}$$

In this case it turns out that

$$\pi_1(\mathcal{G}, Y) = G = \{G_P * G_Q; G_y\}.$$

We have already observed, without proof, that such an amalgamated product $G = A *_U B$ acts on a tree T. This tree can be described as follows. We define first $V(T)$ to be the disjoint union of the set $\{gA \mid g \in G\}$ of left cosets of A in G, and

the set $\{gB \mid g \in G\}$ of left cosets of B in G. $E_+(T)$ is then defined to be the set
$\{gU \mid g \in G\}$ of left cosets of U in G, with $o(gU) = gA$ and $t(gU) = gB$. G acts
on this graph by left multiplication. In order to prove that T is a tree, we need to
prove first that T is connected. The point here is that if, e.g.,

$$f = a_1 b_1 \ldots a_m b_m$$

and

$$g = \alpha_1 \beta_1 \ldots \alpha_n \beta_n,$$

where $a_i, \alpha_j \in A$ and $b_i, \beta_j \in B$, then we have the following path in T from fA to
gB.

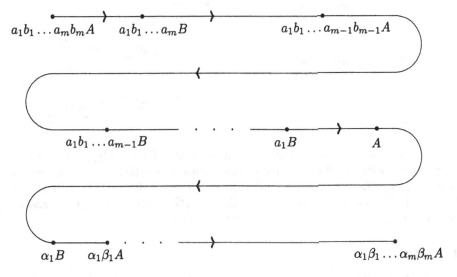

A similar argument shows that there is a path connecting any pair of distinct
vertices in T.

In order to verify that T is a tree, we need to prove that there are no closed paths
in T of length $n > 0$ which do not contain bactrackings. Since T is connected, it
suffices to show that there is no closed path in T of length $n > 0$ beginning at A.
Such a closed path must take the following form (where again $a_i \in A$ and $b_i \in B$):

$$\overset{\bullet}{A} \xrightarrow{\hspace{1cm}} \overset{\bullet}{a_1 B} \xrightarrow{\hspace{1cm}} \overset{\bullet}{a_1 b_1 A} \quad \cdots \quad \overset{\bullet}{} \xrightarrow{\hspace{1cm}} \overset{\bullet}{a_1 b_1 \ldots a_m b_m A}$$

It follows that $a_1 b_1 \ldots a_n b_n \in A$. Consequently, there exists a j with a_j and b_j in
the same factor. Assume, for example, that $a_j \in B$. Then

$$\overset{\bullet}{a_1 b_1 \ldots a_{j-1}} \xrightarrow{\hspace{1cm}} \overset{\bullet}{a_1 b_1 \ldots a_{j-1} b_{j-1} A} \xrightarrow{\hspace{1cm}} \overset{\bullet}{\substack{a_1 b_1 \ldots a_{j-1} b_{j-1} a_j B \\ = a_1 b_1 \ldots a_{j-1} B}}$$

is the desired backtracking.

(3) A tree of groups.

Here $\pi_1(\mathcal{G}, Y)$ turns out to be result of repeatedly forming generalized free products of the vertex groups, with the various edge groups amalgamated and then forming the union of these groups. We shall only consider the special case of an infinite path Y:

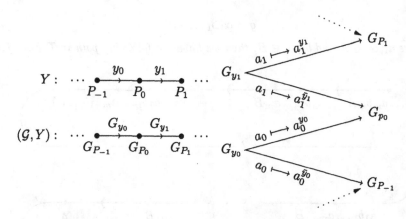

The group $\pi_1(\mathcal{G}, Y)$ is described as being generated by the G_{P_i} with the G_{y_i} identified according to the indicated isomorphisms. In a little more detail, we first form $G_{0,1} = \{G_{P_0} * G_{P_1}; G_{y_1}\}$, and then in succession form $G_{-1,1} = \{G_{P_{-1}} * G_{0,1}; G_{y_0}\}$, $G_{-1,2} = \{G_{-1,1} * G_{P_2}; G_{y_2}\}$, $G_{-2,2} = \{G_{P_{-2}} * G_{-1,2}; G_{y_{-2}}\}, \ldots$, and so on. The group $\pi_1(\mathcal{G}, Y)$ is the union of these groups $G_{-i,i}$.

4 Trees

Lemma 3 *Let X be a graph. Then every tree in X is contained in a maximal tree.*

Proof An inductive limit of trees is again a tree. So it follows, by Zorn's Lemma, that every tree is contained in a maximal one. ∎

Lemma 4 *Let T be a maximal tree in a connected graph X. Then $V(T) = V(X)$.*

Proof Let $v \in V(X)$, $v \notin V(T)$. Let $w \in V(T)$. Then there is a path in X joining v to w. We may assume that v and w are adjacent. Let e be an edge whose extremities are v and w. Since $v \notin V(T)$, $e \notin E(T)$. Now adjoin to T the edges e and $\bar{e}$ and the vertex v. This then defines a subgraph Y of X. Moreover, Y is connected since T is connected and since every closed path in T contains a backtracking, the same is true of Y. So Y is a tree, which properly contains T, a contradiction. ∎

Example 6

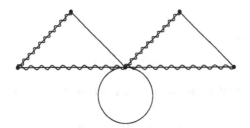

The squiggles outline a maximal tree. Notice that there are usually many such maximal trees.

Now suppose that G acts without inversion on a connected graph $\widetilde{X}$. Then we can form the quotient graph $G\backslash\widetilde{X} = X$, say. Notice that the map

$$p \ : \ \widetilde{X} \longrightarrow X$$

defined by

$$p(v) = \langle v \rangle \quad , \quad p(e) = \langle e \rangle$$

where $v \in V(\widetilde{X})$, $e \in E(\widetilde{X})$, $\langle v \rangle = Gv$, and $\langle e \rangle = Ge$, is a morphism of graphs. Moreover since $\widetilde{X}$ is connected, so is X. Let T be a maximal tree in X. We say that T lifts to a tree $\widetilde{T}$ in $\widetilde{X}$ if $p\,|\,\widetilde{T}$ is an isomorphism between $\widetilde{T}$ and T.

Lemma 5 T *lifts to a tree* $\widetilde{T}$ *in* $\widetilde{X}$.

Proof Let $\widetilde{T}_1$ be a tree in $\widetilde{X}$ which is maximal subject to p mapping $\widetilde{T}_1$ injectively into T. We want to prove that

$$p\left(\widetilde{T}_1\right) = T \ .$$

Suppose the contrary. Then there is a vertex $\langle v \rangle \in V(T)$ such that $\langle v \rangle \notin V(p(\widetilde{T}_1))$. Let $\langle w \rangle \in V(p\left(\widetilde{T}_1\right))$. Then there is a path in T from $\langle w \rangle$ to $\langle v \rangle$. After replacing $\langle w \rangle$ and $\langle v \rangle$ by different vertices if necessary, we may assume that $\langle w \rangle$ and $\langle v \rangle$ are adjacent vertices in T. Let $\langle e \rangle$ be the edge of T with

$$o(\langle e \rangle) = \langle w \rangle \quad , \quad t(\langle e \rangle) = \langle v \rangle \ .$$

Notice that

$$\langle e \rangle = Ge \quad , \quad \langle w \rangle = Gw \quad , \quad \langle v \rangle = Gv \ .$$

In particular, since $\langle w \rangle \in p(\widetilde{T}_1)$, we find that

$$gw \ \in \ \widetilde{T}_1 \qquad \text{for some} \qquad g \in G \ .$$

Consider now the edge

$$ge \in \widetilde{X} .$$

Then $ge \notin \widetilde{T}_1$ because

$$p(ge) = \langle e \rangle .$$

Now adjoin to $\widetilde{T}_1$ the edges ge, $g\,\overline{e}$ and the vertex $t(ge)$. Since

$$p\,t(ge) = t\langle e \rangle = \langle v \rangle,$$

$t(ge) \notin \widetilde{T}_1$. It follows then, as in the proof of Lemma 4, that $\widetilde{T}_2 = \widetilde{T}_1 \cup \{ge, g\,\overline{e}, t(ge)\}$ is a tree and p is injective on $\widetilde{T}_2$. So $p\left(\widetilde{T}_1\right) = T$ after all. ∎

Definition 12 *We sometimes term a lift $\widetilde{T}$ of a maximal tree T in $G\backslash\widetilde{X}$ a tree of representatives (modulo G).*

Notice that every vertex of $\widetilde{X}$ is in one of the subtrees $g\widetilde{T}$ $(g \in G)$, where $\widetilde{T}$ is a tree of representatives modulo G. For if $v \in V(\widetilde{X})$, then $\langle v \rangle \in V(X)$. So $\langle v \rangle \in V(T)$. Hence

$$\langle v \rangle = Gv$$

has a unique preimage gv in $\widetilde{T}$ where $g \in G$. Therefore

$$v \in V(g^{-1}\widetilde{T}) .$$

If $\langle e \rangle \in E(X)$ and $py = \langle e \rangle$ then we sometimes term y a lift of $\langle e \rangle$. Notice that

$$py = Gy = Ge$$

and hence

$$y = ge$$

for some $g \in G$. So the lifts of $\langle e \rangle$ are simply the elements in the G-orbit of e. Notice also that $o(\langle e \rangle) \in V(T)$. Now $o(\langle e \rangle) = Go(e)$, i.e., the lifts of $o(\langle e \rangle)$ comprise the orbit $Go(e)$. Since $o(\langle e \rangle)$ lifts to a vertex of $\widetilde{T}$ it follows that $\langle e \rangle$ lifts to an edge of $\widetilde{X}$ whose origin is in $\widetilde{T}$. We shall have need of such lifts later.

5   The fundamental group of a graph of groups

Let $(\mathcal{G}, Y)$ be a graph of groups. We shall define the fundamental group $\pi_1(\mathcal{G}, Y)$ of $(\mathcal{G}, Y)$ in two ways.

The first way involves a maximal tree T in Y. Suppose $y \in E(T)$. Then we have two monomorphisms

$$\overset{a^{\bar{y}} \longleftarrow a}{\underset{G_{o(y)}}{\bullet} \longleftarrow \underset{G_y}{\bullet}} \qquad \overset{a \longmapsto a^y}{\underset{G_y}{\bullet} \longrightarrow \underset{G_{t(y)}}{\bullet}}$$

of the edge group G_y into the vertex groups of y. We then define G_T to be the group generated by the vertex groups G_P ($P \in V(T)$) with the two images of the edge group G_y in the adjacent vertex groups $G_{o(y)}$ and $G_{t(y)}$ identified according to the prescription

$$a^y = a^{\bar{y}} \qquad (a \in G_y),$$

where here y ranges over all the edges in T. In other words G_T is simply obtained from the vertex groups, by repeatedly forming amalgamated products, where the graph $(\mathcal{G}, Y)$ of groups determines the subgroups to be amalgamated. The group G_T can be described in more precise terms as follows. We choose first a vertex $v_0 \in V(T)$ and define

$$L_0(T) = \{v_0\}.$$

We term $L_0(T)$ the set of vertices at level 0. We then define $L_n(T)$ to consist of those vertices of T which are at a distance n from v_0. Notice that none of the vertices in $L_n(T)$ are the extremities of an edge in T because this would imply the existence of a non-trivial circuit in T. Notice, also, that because T is connected,

$$V(T) = \bigcup_{n=0}^{\infty} L_n(T).$$

We now define G_T as follows. First we define

$$G(0) = G_{v_0}.$$

We then define $G(n+1)$ inductively, assuming that $G(n)$ has already been defined in such a way that it is generated by all of the vertex groups occurring at levels at most n. For each $v \in L_{n+1}(T)$ there is a unique edge $y \in E(T)$ such that $o(y) \in L_n(T)$ with $t(y) = v$. We define $G(v)$ to be the generalised free product of $G(n)$ and G_v with G_y amalgamated according to the monomorphisms of G_y into the vertex groups at its extremities, as dictated by the the graph of groups $(\mathcal{G}, Y)$. $G(n+1)$ is then defined to be the generalised free product of all of these groups $G(v)$ with $G(n)$ amalgamated. Finally we define

$$G_T = \bigcup_{n=0}^{\infty} G(n).$$

It follows that G_T contains an isomorphic copy of each of the vertex groups G_v ($v \in V(T)$ and that

$$\mathrm{gp}(G_{o(y)}, G_{t(y)}) = G_{o(y)} *_{G_y} G_{t(y)}$$

for every edge $y \in E(T)$. It is clear on inspecting the obvious presentation for G_T that this description of G_T does not depend on the choice of v_0. The fundamental group

$$\pi_1(\mathcal{G}, Y, T)$$

of the graph of groups $(\mathcal{G}, Y)$ at T is then defined to be an HNN extension with possibly infinitely many stable letters. The base group is G_T. The choice of the stable letters depends on an orientation, say

$$E(Y) = E_+ \cup E_-$$

of Y. Given such an orientation, for each edge $y \in E_+$, $y \notin E(T)$ we choose a stable letter t_y and define

$$\pi_1(\mathcal{G}, Y, T) = \langle \, G_T \ \{t_y \, | \, y \in E_+ - E(T)\} \mid t_y a^y t_y^{-1}$$
$$= a^{\bar{y}} \, (a \in G_y, y \in E_+ - E(T)) \, \rangle \, .$$

Examples 7

(1) $Y:$ $(\mathcal{G}, Y):$

$$T = \{P\} \quad , \quad G_T = G_P \quad , \quad E_+ = \{y\}$$
$$\pi_1(\mathcal{G}, Y, T) = \langle \, G_t, t_y \, ; \, t_y \, a^y \, t_y^{-1} = a^{\bar{y}} \, \rangle$$

So the fundamental group of a loop of groups is simply an HNN extension with a single stable letter.

(2) $Y:$ $(\mathcal{G}, Y):$

Here T=Y. So

$$\pi_1(\mathcal{G}, Y, T) = G_T = \{G_P * G_Q \, ; \, a^y = a^{\bar{y}} \, (a \in G_y)\} \, .$$

6   The fundamental group of a graph of groups (continued)

The definition of $\pi_1(\mathcal{G}, Y)$ given above seems to depend on the choice of T. There is a somewhat different approach which reveals that this dependence is illusory. This is our next objective.

Suppose then that $(\mathcal{G}, Y)$ is a graph of groups. We define now a group $F(\mathcal{G}, Y)$ whose definition is dictated by the graph of groups $(\mathcal{G}, Y)$. To this end let us choose an orientation $E(Y) = E_+(Y) \cup E_-(Y)$ of Y. Then $F(\mathcal{G}, Y)$ is an HNN extension, where the base group is the free product of all of the vertex groups G_p:

$$F(\mathcal{G}, Y) = \langle \, (\prod_{P \in V(Y)}^{\star} G_P) \cup E_+(Y) \, ; \; y a^y y^{-1} = a^{\bar{y}} \; (a \in G_y, \; y \in E_+(Y)) \, \rangle \, .$$

We now interpret $\displaystyle\prod_{P \in V(Y)}^{\star} G_P$ as being given by generators and defining relations. Then $F(\mathcal{G}, Y)$ can be presented in the form:

$$F(\mathcal{G}, Y) = \langle \, (\prod_{P \in V(Y)}^{\star} G_P) \cup E(Y) \, ; \; \bar{y} = y^{-1}, \, y a^y y^{-1} = a^{\bar{y}} \; (a \in G_y, \; y \in E(Y)) \, \rangle \, .$$

It is easy to obtain a presentation from $F(\mathcal{G}, Y)$ for

$$\pi_1(\mathcal{G}, Y, T) \, ,$$

where T is a maximal tree in Y. Indeed define

$$Q(\mathcal{G}, Y, T) = \langle \, F(\mathcal{G}, Y) \mid y = 1 \; if \; y \in E(T) \, \rangle \, .$$

In other words, we add to $F(\mathcal{G}, Y)$ the relations $y = 1$, for every edge y of T. Now denote the canonical image of $y \in E(Y)$ by t_y. Then

$$t_y a^y t_y^{-1} = a^{\bar{y}} \; (a \in G_y).$$

Since $t_y = 1$ if $y \in E(T)$, it follows that

$$a^y = a^{\bar{y}} \; (a \in G_y, y \in E(T)).$$

It follows from these remarks that

$$Q(\mathcal{G}, Y, T) \cong \pi_1(\mathcal{G}, Y, T) \, .$$

We are actually more interested in a particular *subgroup* of $F(\mathcal{G}, T)$, which we will compare with $Q(\mathcal{G}, Y, T)$. To this end let P_0 be a chosen vertex in Y and let c be a path in Y with origin P_0. As usual we think of c as a sequence of edges

$$y_1, \ldots, y_n .$$

Put $P_i = t(y_i)$ $(i = 1, \ldots, n)$. Notice that $o(y_{i+1}) = t(y_i)$ $(i = 1, \ldots, n-1)$.

Definition 13 *A word of type c in $F(\mathcal{G}, Y)$ is a pair (c, μ) where*

$$\mu = (r_0, \ldots, r_n)$$

is a sequence of elements

$$r_i \in G_{P_i} \ (i = 0, \ldots, n).$$

Definition 14 *Let (c, μ) be a word of type c. Then we define*

$$|c, \mu| = r_0 y_1 r_1 y_2 \ldots y_n r_n \qquad (\in F(\mathcal{G}, Y))$$

and say that $|c, \mu|$ is the element or word in $F(\mathcal{G}, Y)$ associated with (c, μ). Notice that when $n = 0$, $|c, \mu| = r_0$.

Definition 15 *Let*

$$\pi_1(\mathcal{G}, Y, P_0) = \{ \, |c, \mu| \ \mid \ c \text{ is a closed path in } Y \text{ beginning (and ending) at } P_0 \, \} \, .$$

It is clear that $\pi_1(\mathcal{G}, Y, P_0)$ is a subgroup of $F(\mathcal{G}, Y)$.

Proposition 1 $\pi_1(\mathcal{G}, Y, P_0) \cong \pi_1(\mathcal{G}, Y, T)$.

Proof It suffices, by the remarks above, to prove that

$$\pi_1(\mathcal{G}, Y, P_0) \cong Q(\mathcal{G}, Y, T) \, .$$

We first concoct a homomorphism

$$f \, : \, Q(\mathcal{G}, Y, T) \longrightarrow \pi_1(\mathcal{G}, Y, P_0) \, .$$

In order to do so, suppose that $P \in V(Y)$. Then there is a unique geodesic c_P in T joining P_0 to P:

$$c_P = y_1, \ldots, y_n \, .$$

Put

$$\gamma_P = y_1 \ldots y_n \in F(\mathcal{G}, Y) \,.$$

Then for each $x \in G_P$ define

$$x' = \gamma_P \, x \, \gamma_P^{-1}$$

and for each edge $y \in E(Y)$ define

$$y' = \gamma_{o(y)} y \gamma_{t(y)}^{-1} \,.$$

Notice that $x', y' \in \pi_1(\mathcal{G}, Y, P_0)$.

If $y \in E(T)$, then $y' = 1$. Furthermore, if $y \in E(Y)$,

$$\overline{y}' y' = \gamma_{o(\overline{y})} \overline{y} \gamma_{t(\overline{y})}^{-1} \cdot \gamma_{o(y)} y \gamma_{t(y)}^{-1} = 1 \,.$$

And if $a \in G_y$,

$$\begin{aligned}
y'(a^y)'y'^{-1} &= \gamma_{o(y)} y \gamma_{t(y)}^{-1} \cdot \gamma_{t(y)} a^y \gamma_{t(y)}^{-1} \cdot \gamma_{t(y)} y^{-1} \gamma_{o(y)}^{-1} \\
&= \gamma_{o(y)} y a^y y^{-1} \gamma_{o(y)}^{-1} \\
&= \gamma_{o(y)} a^{\overline{y}} \gamma_{o(y)}^{-1} \\
&= \left(a^{\overline{y}}\right)' \,.
\end{aligned}$$

We now map a set of generators of $\pi_1(\mathcal{G}, Y, T)$ to $\pi_1(\mathcal{G}, Y, P_0)$ as follows

$$\begin{aligned}
x &\longmapsto x' \qquad (x \in G_P, \, P \in V(Y)), \\
t_y &\longmapsto y' \qquad (y \in E_+ - E(T)).
\end{aligned}$$

Notice that the images of the elements x and the t_y under this mapping satisfy the defining relations of $\pi_1(\mathcal{G}, Y, T)$. So, by Dyck's Theorem, the given mapping defines a homomorphism

$$h \,:\, \pi_1(\mathcal{G}, Y, T) \longrightarrow \pi_1(\mathcal{G}, Y, P_0) \,.$$

Consequently, using the canonical isomorphism between

$$\pi_1(\mathcal{G}, Y, T) \quad \text{and} \quad Q(\mathcal{G}, Y, T),$$

we obtain a homomorphism

$$f \,:\, Q(\mathcal{G}, Y, T) \longrightarrow \pi_1(\mathcal{G}, Y, P_0) \,.$$

Let now p be the canonical projection of $F(\mathcal{G}, Y)$ onto $Q(\mathcal{G}, Y, T)$ and let i be the restriction of p to $\pi_1(\mathcal{G}, Y, P_0)$:

$$i = p \,\big|\, \pi_1(\mathcal{G}, Y, P_0) \,.$$

Observe that if $y \in E(T)$, then
$$i(y) = 1 .$$

Hence
$$i(\gamma_P) = 1 \qquad \text{for all} \qquad P \in V(Y) .$$

Consequently
$$i(x') = x \; (x \in G_P, P \in V(Y)), \; i(y') = y \; (y \in E_+ - E(T)).$$

Thus
$$i \circ f = 1 . \tag{1}$$

In order to complete the proof of the proposition it suffices to prove that
$$f \circ i = 1 . \tag{2}$$

Suppose then that c is a closed path with origin P_0, (c, μ) a word of type c and that
$$|c, \mu| = r_0 y_1 r_1 \ldots y_n r_n .$$

Now
$$r'_i = \gamma_{P_i} r_i \gamma_{P_i}^{-1} \; (i = 0, \ldots, n), \; y'_i = \gamma_{P_{i-1}} y_i \gamma_{P_i}^{-1} \; (i = 1, \ldots, n).$$

Therefore
$$r'_0 y'_1 r'_1 \ldots y'_n r'_n = \gamma_{P_0} (r_0 y_1 r_1 \ldots y_n r_n) \gamma_{P_0}^{-1} .$$

Since $\gamma_{P_0} = 1$ by definition,
$$r'_0 y'_1 r'_1 y'_n r'_n = r_0 y_1 r_1 \ldots y_n r_n,$$

i.e., (2) holds. This completes the proof. ∎

Corollary 1 *The groups $\pi_1(\mathcal{G}, Y, T)$, $\pi_1(\mathcal{G}, Y, P_0)$ are isomorphic, hence independent of either the choice of T or P_0.*

Proof Let P_0, $\overline{P}_0$ be arbitrarily chosen and fix T. Then
$$\pi_1(\mathcal{G}, Y, P_0) \cong \pi_1(\mathcal{G}, Y, T) \cong \pi(\mathcal{G}, Y, \overline{P}_0) .$$

(In fact $\pi_1(\mathcal{G}, Y, P_0)$ and $\pi_1(\mathcal{G}, Y, \overline{P}_0)$ are conjugate in $F(\mathcal{G}, Y)$). But if $\overline{T}$ is now any other maximal tree in Y, we again find that
$$\pi_1(\mathcal{G}, Y, P_0) \cong \pi_1(\mathcal{G}, Y, \overline{T}) .$$

This completes the proof. ∎

One more remark, before we move on to determine some graphs of groups.

Suppose $(\mathcal{G}, Y)$ is a graph of groups, $y \in E(Y)$. Then we denote the image of G_y in $G_{t(y)}$ under the map

$$a \longmapsto a^y \qquad (a \in G_y)$$

by G_y^y.

Definition 16 *Let (c, μ) be a word of type c where c is a closed path in Y beginning at P_0 with edges $y_1, \ldots, y_n$. Suppose that*

$$\mu = (r_0, \ldots, r_n) .$$

Then we term (c, μ) a reduced word if
(i) $r_0 \neq 1$ if $n = 0$;
(ii) $r_i \notin G_{\bar{y}_i}^{y_i}$ whenever $y_{i+1} = \bar{y}_i$ $(i = 1, \ldots, n-1)$.

Now $F(\mathcal{G}, Y)$ is an HNN extension of the free product of the groups G_P. It follows immediately from Britton's Lemma that

Proposition 2 *If (c, μ) is a reduced word, then $|c, \mu| \neq 1$.*

7   Group actions and graphs of groups

Let G be a group acting without inversion on a connected graph $\widetilde{Y}$. The key to understanding the structure of a group acting without inversion on a tree is the fact that we can associate to this action of G on $\widetilde{Y}$, a graph $(\mathcal{G}, Y)$ of groups, where $Y = G \backslash \widetilde{Y}$. To this end, let T be a maximal tree in Y, $\widetilde{T}$ a lift of T in $\widetilde{Y}$. Let

$$p : \widetilde{Y} \longrightarrow Y$$

be the canonical projection. Then, restricting p to T, we can view $p^{-1}|T$ as an isomorphism from T to $\widetilde{T}$. Our objective is to extend this isomorphism to a map

$$q : Y \longrightarrow \widetilde{Y}.$$

It is important to note that q need not be a morphism of graphs, even though $p^{-1}|T$ is such a morphism.

Now q is already defined on T, hence on all of the vertices of Y. We need to define q on the edges y of Y. If $y \in E(T)$, then qy has already been defined. We are left with the edges $y \notin E(T)$. Choose an orientation $E(Y) = E_+(Y) \cup E_-(Y)$ of Y. We will define q on the positive edges y not in T and extend q to the other edges by defining $q\overline{y} = \overline{(qy)}$. Let y be a positive edge not in T. Then there exists an edge $\widetilde{y} \in E(\widetilde{Y})$, which is a lift of y, with

$$o(\widetilde{y}) \in V(\widetilde{T}).$$

Define

$$qy = \widetilde{y}.$$

This completes the definition of q. Notice that

$$q\overline{y} = \overline{(qy)} \text{ for all } y \in E(Y).$$

Put

$$e(y) = \begin{cases} 0 & \text{if } y \in E_+, \\ 1 & \text{if } y \notin E_+. \end{cases}$$

Notice that

$$p\,t(\widetilde{y}) = t(y).$$

So $t(\widetilde{y})$ and $q\,t(y)$ are both lifts of $t(y)$. Hence there exists an element $g_y \in G$ such that

$$t(\widetilde{y}) = g_y\, q\,t(y).$$

Furthermore, we define

$$g_{\overline{y}} = g_y^{-1} \text{ if } y \in E_-, \ g_y = 1 \text{ if } y \in E(T).$$

It follows that

$$g_{\overline{y}} = g_y^{-1} \ (y \in E(Y)) \text{ and } g_y = 1 \text{ if } y \in E(T).$$

In addition, for each $y \in E(Y)$

$$o(qy) = g_y^{-e(y)} qo(y)$$
$$t(qy) = g_y^{1-e(y)} qt(y).$$

We are now in a position to define the graph $(\mathcal{G}, Y)$ of groups. The edge and vertex groups of $(\mathcal{G}, Y)$ are the stabilizers of the images of vertices and edges of Y under q. In detail, then, we define the vertex and edge groups of $(\mathcal{G}, Y)$ as follows:

$$\mathcal{G}(P) = G_{qP} = \{g \in G \mid g(qP) = qP\} \stackrel{\text{def}}{=} G_P \ (P \in V(Y));$$
$$\mathcal{G}(y) = G_{qy} = \{g \in G \mid g(qy) = qy\} \stackrel{\text{def}}{=} G_y \ (y \in E(Y)).$$

We now define the edge monomorphisms

$$a \longmapsto a^y \qquad (a \in G_y, y \in E(Y))$$

by putting

$$a^y = g_y^{e(y)-1} a g_y^{1-e(y)} \ (y \in E(Y)).$$

We have two things to check. First

$$G_y = G_{qy} = G_{\overline{qy}} = G_{q\overline{y}} = G_{\overline{y}}.$$

Second, we have to make sure that the mapping $a \longmapsto a^y$ is a monomorphism from G_y into $G_{t(y)}$, i.e., from G_{qy} into $G_{qt(y)}$. Now $qt(y) = g^{e(y)-1}t(qy)$ and $G_{qy} \leq G_{t(qy)}$. Consequently $g_y^{e(y)-1} a g_y^{1-e(y)} \in G_{qt(y)}$. Therefore the mapping $a \longmapsto a^y$ is a monomorphism from G_y into $G_{t(y)}$, as desired.

Since T plays a role here in the definition of $(\mathcal{G}, Y)$ we sometimes denote $(\mathcal{G}, Y)$ by $(\mathcal{G}, Y, T)$ and refer to $(\mathcal{G}, Y, T)$ as the graph of groups $(\mathcal{G}, Y)$ at T.

The following theorem holds.

Theorem 2 *Let G be a group acting without inversion on a tree $\widetilde{Y}$. Then*

$$\pi_1(\mathcal{G}, Y, T) \cong G,$$

where $Y = G\backslash\widetilde{T}$ and $(\mathcal{G}, Y, T)$ is the graph of groups associated to the action of G on $\widetilde{Y}$ at the maximal tree T of Y.

The connection between $\pi_1(\mathcal{G}, Y, T)$ and G is the natural one, coming out of the very definition of $(\mathcal{G}, Y)$. To begin with we concoct a homomorphism from $\pi_1(\mathcal{G}, Y, T)$ onto G. Now $\pi_1(\mathcal{G}, Y, T)$ is made up from the vertex groups of $(\mathcal{G}, Y)$, together with a set of stable letters t_y, one for each positive edge $y \notin E(T)$. Each of the vertex groups is a subgroup of G, namely the stabilizer G_{qP} of the vertex qP in $\widetilde{T}$. We have arranged that the relations defining $\pi_1(\mathcal{G}, Y, T)$ mimic those that hold between the corresponding subgroups of G. Therefore there is a canonical homomorphism from $\pi_1(\mathcal{G}, Y, T)$ to G, which is the identity on the G_{qP}, and maps each t_y to g_y:

$$\Theta : \pi_1(\mathcal{G}, Y, T) \longrightarrow G$$

Our first step in the proof of Theorem 2 is to prove that Θ is onto. This is taken care of by the following

Lemma 6 *Let G be a group acting without inversion on a connected graph X, U a tree of representatives of X modulo G and Z a subgraph of X such that*
(i) if $z \in E(Z)$ then either $o(z) \in V(U)$ or $t(z) \in V(U)$;
(ii) $GZ = X$.
If for each $z \in E(Z)$ with $o(z) \in V(U)$, $g_z \in G$ is chosen so that $g_z^{-1} t(z) \in V(U)$, then

$$G = \text{gp}\left(\{ G_P \mid P \in V(U) \} \cup \{ g_z \mid z \in E(Z), o(z) \in V(U) \}\right) \qquad (3)$$

where as usual

$$G_P = \{ g \in G \mid gP = P \} .$$

To see how Lemma 6 implies that Θ is onto, we take $X = \widetilde{Y}$, $U = \widetilde{T}$, $Y = G\backslash\widetilde{Y}$, $Z = qY$ and the g_z to be the g_y.

Proof of Lemma 6. Let us denote the right-hand-side of (3) by H. Our objective is to prove that $H = G$.

Let $p : X \longrightarrow G\backslash X$ be the canonical projection. By definition pU is a maximal tree in $G\backslash X$ and hence contains all of the vertices of $G\backslash X$. So, as we have already noted previously,

$$V(X) = V(GU) .$$

Suppose that $V(HU) = V(GU)$. Let $g \in G$, $P \in V(U)$. Then $gP \in V(HU)$. So there exists $Q \in V(U)$, $h \in H$ such that $gP = hQ$. Since p is injective on U, this implies $Q = P$. Hence

$$h^{-1}g \in G_P .$$

So by the very definition of H, $h^{-1}g \in H$ and hence so does $h \cdot h^{-1}g = g$. It suffices, therefore, to prove that $V(HU) = V(X)$. Suppose the contrary. This means that there is a vertex Q of X which is not in any of the subtrees hU ($h \in H$). Let Q be chosen to be as "close" to one of these subtrees as possible, $Q \notin V(HU)$. Since X is connected, we can assume that there is an edge $y \in E(X)$ such that

$$P = o(y) \in V(HU) \quad , \quad t(y) = Q .$$

On replacing y by hy, where h is a suitably chosen element of H, we can assume that $P \in V(U)$. Now

$$GZ = X .$$

So there exists $f \in G$ such that

$$z = fy \in E(Z) .$$

We then have two possibilities.

(1) $o(z) \in V(U)$. Notice that $o(z) = fP$ and $P \in V(U)$. Again using the injectivity of p we deduce that $fP = P$, i.e., $f \in G_P \le H$. By *(iii)* of the hypothesis,

$$\left(g_z^{-1}f\right)Q \in V(U) ,$$

where $g_z \in H$. Hence $Q \in V(HU)$ after all.

(2) $t(z) \in V(U)$. There exists again g_z such that

$$g_z^{-1}o(z) \in V(U) .$$

Now $o(z) = fP$. So $g_z^{-1}fP \in V(U)$. Since $P \in V(U)$ this implies as before that $g_z^{-1}f \in G_P \le H$. This means that $f \in H$. But

$$t(z) = fQ \in V(U) .$$

Therefore

$$Q \in V(HU)$$

once again. This completes the proof.

We have proved the surjectivity of Θ. We still need to prove that Θ is a monomorphism. This will be accomplished by first proving Theorem 3 in **8**; the proof is then completed in **9**.

8   Universal covers

Let $(\mathcal{G}, Y)$ be a graph of groups, T a maximal tree in Y and

$$G = \pi_1(\mathcal{G}, Y, T) \ .$$

Suppose that $\widetilde{Y}$ is a tree, that G acts on $\widetilde{Y}$, that

$$G\backslash\widetilde{Y} \xrightarrow{\sim} Y$$

and that the graph of groups that we associate to this action of G on $\widetilde{Y}$ is (isomorphic, in the obvious sense) to $\mathcal{G}(Y, T)$. Suppose that we concoct a map, as before, $q : Y \longrightarrow \widetilde{Y}$ which is an injective morphism on T. Now $GV(qT) = V(\widetilde{Y})$ and $G_{qP} = G_P$; this implies that $V(\widetilde{Y})$ can be identified with the disjoint union of all of the left cosets G/G_P $(P \in V(Y))$ of G_P in G. Furthermore $GqE(Y) = E(\widetilde{Y})$ and $G_{qy} = G_y$. So $E(\widetilde{Y})$ can be identified with the disjoint union of the set of all the left cosets G/G_y $(y \in E(Y))$ of G_y in G.

Our aim now, given a graph $(\mathcal{G}, Y)$ of groups, a maximal tree T in Y and the fundamental group $G = \pi_1(\mathcal{G}, Y, T)$ of this graph of groups, is to reverse the process described above. That is, to construct a *tree* $\widetilde{Y} = \widetilde{Y}(\mathcal{G}, Y, T)$ and an action of G on $\widetilde{Y}$ so that the graph of groups associated to this action is (isomorphic to) $(\mathcal{G}, Y)$.

To this end, let $E(Y) = E_+(Y) \cup E_-(Y)$ be the usual orientation of Y. For each $y \in E_+(Y)$, define

$$G(y) = G_{\overline{y}}^{\overline{y}} \qquad G(\overline{y}) = G(y).$$

Notice that

$$G(y) \leq G_{t(\overline{y})} = G_{o(y)} \ (y \in E_+(Y)).$$

We are now in a position to *define* $\widetilde{Y}$.

(i) We define $V(\widetilde{Y})$ to be the disjoint union of all of the right cosets gG_P of G_P in G, where P ranges over all of the vertices of Y. We emphasise that two such cosets gG_P and hG_Q are, by definition, disjoint if $P \neq Q$.

(ii) $E_+\left(\widetilde{Y}\right)$ is the disjoint union of the right cosets $gG(y)$ of the $G(y)$ in G, where here y ranges over $E_+(Y)$. We then define $E_-\left(\widetilde{Y}\right)$ to be the disjoint union of the right cosets $gG(y)$ of the $G(y)$ in G, where here y ranges over $E_-(Y)$. $E\left(\widetilde{Y}\right)$ is then taken to be the disjoint union of $E_+\left(\widetilde{Y}\right)$ and $E_-\left(\widetilde{Y}\right)$. It is important here to observe that if we view the cosets $gG(y)$ and $hG(z)$ as *edges* then they are equal only if $y = z$ and the sets $gG(y)$, $hG(y)$ coincide. So, in particular, the *edges* $gG(y)$ and $gG(\overline{y})$ are distinct, notwithstanding the fact that they are set-theoretically equal.

(iii) If $y \in E(Y)$, then we define

$$o\left(gG(y)\right) = gt_y^{-e(y)}G_{o(y)} \quad , \quad t\left(gG(y)\right) = g\,t_y^{1-e(y)}\,G_{t(y)},$$

where, as before, $e(y) = 0$ if $y \in E_+(Y)$ and $e(y) = 1$, otherwise. Notice that if $y \in E_+$, then $o(gG(y)) = gG_{o(y)}$ and $t(gG(y)) = gt_yG_{t(y)}$; and if $y \in E_-$, then

$$o(gG(y)) = gt_y^{-1}G_{o(y)} = gt_{\overline{y}}G_{t(\overline{y})} = t(gG(\overline{y}))$$

and

$$t(gG(y)) = gG_{t(y)} = gG_{o(\overline{y})} = o(gG(\overline{y})).$$

Finally, we define

$$\overline{gG(y)} = gG(\overline{y});$$

we emphasise that here $gG(y)$ and $gG(\overline{y})$ are edges and therefore they are, by definition, distinct.

Definition 17 *The graph $\widetilde{Y}$ is called the universal covering of Y relative to the graph of groups $(\mathcal{G}, Y)$ at T.*

Notice that G acts without inversion on $\widetilde{Y}$ in the obvious way, by left multiplication. Notice also that

$$V\left(G\backslash\widetilde{Y}\right) = \{\, \langle G_P \rangle \mid P \in V(Y) \,\}$$
$$E_+\left(G\backslash\widetilde{Y}\right) = \{\, \langle G(y) \rangle \mid y \in E_+(Y) \,\}$$

Thus the maps

$$\langle G_P \rangle \longmapsto P \quad , \quad \langle G(y) \rangle \longmapsto y$$

define a morphism of graphs – indeed an isomorphism

$$G\backslash\widetilde{Y} \xrightarrow{\sim} Y.$$

Our main objective is to prove the following

Theorem 3 $\widetilde{Y}$ *is a tree.*

Proof There are two steps in the proof. The first is to prove that $\widetilde{Y}$ is connected. This follows, as we will shortly see, from the fact that G is generated by the G_P together with the t_y. The second, is to prove that every closed path in $\widetilde{Y}$ of length $n > 0$ contains a backtracking. It turns out that every such closed path corresponds to a word w in the elements of the G_P and the t_y with $w = 1$. Britton's Lemma and the normal form theorem for amalgamated products can then be used to show that w has a special form, which in turn implies that the given path has a backtracking. In more detail, then, we first prove connectedness.

(*i*) Notice that T lifts to a tree $\widetilde{T}$ in $\widetilde{Y}$ as follows. We define $V(\widetilde{T})$ to consist of the vertices $\widetilde{P} = G_P$, where $P \in V(Y)$, and $E(\widetilde{T})$ to consist of the edges $\widetilde{y} = G(y)$, where $y \in E(T)$. Observe that if $P, Q \in V(T)$, then there is a path $y_1, \ldots, y_n$ in T from P to Q. Hence

$$\widetilde{y_1} = G(y_1), \ldots, \widetilde{y_n} = G(y_n)$$

is a path in $\widetilde{T}$ from $\widetilde{P}$ to $\widetilde{Q}$. In order to prove connectedness, we need to prove that if $g\widetilde{P}$ and $h\widetilde{Q}$ are any pair of vertices in $\widetilde{Y}$, then there is a path in $\widetilde{Y}$ from $g\widetilde{P}$ to $h\widetilde{Q}$. In view of the fact that G acts on $\widetilde{Y}$, it suffices to prove that there is a path in $\widetilde{X}$ from $\widetilde{P}$ to $f\widetilde{Q}$, where $f \in G$. Now

$$f = a_1 \ldots a_m,$$

where a_i either belongs to a G_R or is a t_y and $y \in E(Y)$, $y \notin E(T)$. Consider first the case $m = 1$. Suppose that $a_1 \in G_R$. There is a path in T from R to Q, say $z_1, \ldots, z_l$. Then we have an edge

$$
\begin{array}{cc}
& a_1 G(z_1) \\
\bullet\!\!\!\!\!\!\!\!&\xrightarrow{\hspace{2cm}}\!\!\!\!\!\!\!\bullet \\
a_1 \widetilde{R} = \widetilde{R} & a_1 t(z_1)
\end{array}
$$

from $\widetilde{R}$ to $a_1\widetilde{t(z_1)}$. Now $a_1\widetilde{t(z_1)} \in a_1\widetilde{T}$. So there is a path in the tree $a_1\widetilde{T}$ from $a_1\widetilde{t(z_1)}$ to $a_1\widetilde{Q}$. But there is a path in $\widetilde{T}$ from $\widetilde{P}$ to $\widetilde{R}$. Therefore there is a path in $\widetilde{Y}$ from $\widetilde{P}$ to $a_1\widetilde{Q}$.

If $a_1 = t_y$, where y is a positive edge, not in T, then the edge $G(y)$ takes the form

$$
\begin{array}{cc}
& G(y) \\
\bullet\!\!\!\!\!\!\!\!&\xrightarrow{\hspace{2cm}}\!\!\!\!\!\!\!\bullet \\
G_{o(y)} & t_y G_{o(y)}
\end{array}
$$

So we have an edge with origin in $\widetilde{T}$ and terminus in $t_y\widetilde{T}$. Thus, as before, there is a path from $\widetilde{P}$ to $a_1\widetilde{Q}$. Finally if $a = t_{\overline{y}}$, then we have an edge from $G_{t(y)}$ to

$t_{\widetilde{y}}G_{o(y)}$ and the same kind of argument as before applies. The case $m > 1$ is now an easy consequence of these remarks. For we have a path from $\widetilde{P}$ to $a_1\widetilde{P}$ in $\widetilde{Y}$, and inductively a path from $\widetilde{P}$ to $a_2\ldots a_m\widetilde{Q}$. Hence there is a path from $a_1\widetilde{P}$ to $a_1a_2\ldots a_m\widetilde{Q}$, and consequently also a path from $\widetilde{P}$ to $f\widetilde{Q}$, as needed.

(*ii*) It remains to prove that $\widetilde{Y}$ is a tree. Thus we have to prove that every closed path of length $n > 0$ in $\widetilde{Y}$ contains a backtracking. To this end, suppose that

$$s_1\widetilde{y_1} = s_1G(y_1),\ldots,s_n\widetilde{y_n} = s_nG(y_n)$$

is such a closed path. This path projects onto the closed path

$$c = y_1,\ldots,y_n$$

in Y. So if $o(y_i) = P_i$, then $t(y_n) = P_0$. Put

$$e_i = e(y_i), t_i = t_{y_i} \ (i = 1,\ldots,n).$$

Then

$$t(s_n\widetilde{y_n}) = s_nt_n^{1-e_n}\widetilde{P_0} = s_1t_1^{-e_1}\widetilde{P_0} = o(s_1\widetilde{y_1})$$
$$t(s_1\widetilde{y_1}) = s_1t_1^{1-e_1}\widetilde{P_1} = s_2t_2^{-e_2}\widetilde{P_1} = o(s_2\widetilde{y_2})$$

$$\vdots$$

$$t(s_{n-1}\widetilde{y_{n-1}}) = s_{n-1}t_{n-1}^{1-e_{n-1}}\widetilde{P_{n-1}} = s_nt_n^{-e_n}\widetilde{P_{n-1}} = o(s_n\widetilde{y_n})$$

Put $q_i = s_it_i^{-e_i}$. Then

$$q_nt_nr_n = q_1 \ (r_0 \in G_{P_0})$$
$$q_1t_1r_1 = q_2 \ (r_1 \in G_{P_1})$$

$$\vdots$$

$$q_{n-1}t_{n-1}r_{n-1} = q_n \ (r_{n-1} \in G_{P_{n-1}})$$

It follows that

$$t_1r_1 = q_1^{-1}q_2, \ t_2r_2 = q_2^{-1}q_3, \ t_nr_n = q_n^{-1}q_1,$$

and so, on multiplying these terms together, we find that

$$t_1r_1\ldots t_nr_n = 1.$$

Let (c, μ) be the word of type c defined by

$$\mu = (1, r_1,\ldots,r_n).$$

We claim that μ is reduced, i.e., if $y_{i+1} = \overline{y}_i$, then $r_i \notin G_{y_i}^{y_i}$. Suppose that $y_{i+1} = \overline{y}_i$; then $t_{i+1} = t_i^{-1}$ and $e_{i+1} = 1 - e_i$. Our objective is to prove that

$$r_i \notin G_{y_i}^{y_i}.$$

Suppose the contrary. It follows from the relation $q_i t_i r_i = q_{i+1}$ that $s_i t_i^{-e_i} t_i r_i = s_{i+1} t_{i+1}^{-e_{i+1}}$. Therefore

$$s_i^{-1} s_{i+1} \in t_i^{1-e_i} G_{y_i}^{y_i} t_i^{e_i-1}.$$

If $y_i \in T$, then $t_i = 1$ and

$$t_i^{1-e_i} G_{y_i}^{y_i} t_i^{e_i-1} = G_{y_i}^{y_i} = G_{\overline{y}_i}^{\overline{y}_i}.$$

If $y_i \in E_+(Y)$, $y_i \notin T$, then $e_i = 0$ and again $t_i^{1-e_i} G_{y_i}^{y_i} t_i^{e_i-1} = G_{\overline{y}_i}^{\overline{y}_i}$. Similarly $t_i^{1-e_i} G_{y_i}^{y_i} t_i^{e_i-1} = G_{\overline{y}_i}^{\overline{y}_i}$ if $y_i \in E_-(Y)$. Therefore

$$s_i^{-1} s_{i+1} \in G_{\overline{y}_i}^{\overline{y}_i}.$$

We now go back to the path $s_1\widetilde{y}_1, \ldots, s_n\widetilde{y}_n$. Consider the consecutive edges $s_i\widetilde{y}_i = s_i G(y_i)$, $s_{i+1}\widetilde{y}_{i+1} = s_{i+1} G(y_{i+1})$ and observe that

$$\overline{s_{i+1}\widetilde{y}_{i+1}} = \overline{s_{i+1}G(y_{i+1})} = s_{i+1}G(\overline{y_{i+1}}) = s_i G(y_i) = s_i\widetilde{y}_i.$$

This implies that there is a backtracking in the path $s_1\widetilde{y}_1, \ldots, s_n\widetilde{y}_n$, which is not the case. Hence $r_i \notin G_{y_i}^{y_i}$, as claimed. Consequently (c, μ) is reduced and therefore

$$|c, \mu| = 1.t_1 r_1 t_2 \ldots t_n r_n \neq 1,$$

a contradiction. Thus there is no closed path of length $n > 0$ in $\widetilde{Y}$ without a backtracking. This completes the proof of Theorem 3.

We illustrate Theorem 3 by describing a few examples.

Examples 8 *(1)* $(\mathcal{G}, Y)$ *a loop of groups.*

$$G = \pi_1(\mathcal{G}, Y, T) = \langle\, G_P, t_y\ ;\ t_y a^y t_y^{-1} = a^{\overline{y}}\,\rangle.$$
$$V\!\left(\widetilde{Y}\right) = G/G_P = \{\, gG_P \mid g \in G \,\}$$
$$E_+\!\left(\widetilde{Y}\right) = G/G_y = \{\, gG_y \mid g \in G \,\}.$$

Thus the positive edges of $\widetilde{Y}$ all have the form

$$
\xrightarrow[gG_P]{\quad gG_y \quad} gt_yG_P
$$

It is instructive to try to see why $\widetilde{Y}$ is connected. The thing to notice is that if $g \in G$, then

$$
g = g_0\, t_y^{\varepsilon_1}\, g_1\, \ldots\, t_y^{\varepsilon_n}\, g_n
$$

where

$$
g_i \in G_P \quad (i = 0,\ldots,n)\ ,\quad \varepsilon_i = \pm 1 \quad (i = 1,\ldots,n)\ .
$$

One wants to find a path, e.g., from

$$
G_P \qquad to \qquad gG_P\ .
$$

Let's look at the case $n = 1$, $\varepsilon_1 = 1$: we have the edge

$$
\xrightarrow[\substack{gG_P \\ \| \\ g_0G_P}]{\quad gG_y \quad} \substack{g_0t_yg_1G_P \\ \| \\ g_0t_ygP}
$$

(2) $(\mathcal{G},Y)$ a segment of groups.

Here any maximal tree T coincides with Y. We have already described the universal covering $\widetilde{Y}$ of this segment of groups at T. We remind the reader that if

$$
Y: \quad \xrightarrow[P]{\quad y \quad} Q
$$

then $V\left(\widetilde{Y}\right)$ is the disjoint union

$$
V\left(\widetilde{Y}\right) = G/G_P \bigcup G/G_Q
$$

and

$$
E_+\left(\widetilde{Y}\right) = G/G_y
$$

Observe that the positive edges of $\widetilde{Y}$ all take the form

$$
Y: \quad \xrightarrow[\substack{gG_P}]{\quad gGy \quad} gG_Q
$$

9   The proof of Theorem 2

In order to prove Theorem 2, we make use of the existence of a universal covering. Suppose that G acts without inversion on a tree X and that $Y = G\backslash X$. Form the corresponding graph of groups $(\mathcal{G}, Y)$ associated to this action of G on X relative to the choice of a maximal tree T in Y. Now form

$$H = \pi_1(\mathcal{G}, Y, T) \ .$$

As we noted before there is an epimorphism

$$\Theta : H \longrightarrow G \ .$$

Let $\widetilde{Y}$ be the universal covering of Y corresponding to this graph of groups $(\mathcal{G}, Y)$ relative to T. Now the definition of $(\mathcal{G}, Y)$ involves the use of a map

$$q : Y \longrightarrow X \ .$$

As we have already observed

$$V(X) = \{\, g\, q(P) \mid g \in G,\, P \in V(Y)\,\} \ .$$

Define a map φ on $V\left(\widetilde{Y}\right)$ by

$$\varphi : hG_P \longmapsto \Theta(h)\, q(P) \qquad (h \in H) \ .$$

Notice that since we are using H to denote $\pi_1(\mathcal{G}, Y, T)$ the G_P are now subgroups of H. It is clear that φ is surjective. Similarly, notice that

$$E_+(X) = \{\, g\, q(y) \mid g \in G,\, y \in E_+(Y)\,\} \ .$$

Define a map, again denoted φ, on $E_+\left(\widetilde{Y}\right)$ by

$$\varphi : hG_y \longmapsto \Theta(h)q(y) \ .$$

It is easy to see that φ defines a surjective morphism of graphs. Notice also that Θ induces an isomorphism between the stabilisers of the edges and the vertices. It follows that Θ is *locally injective*, i.e., it is injective on the star of any vertex. These stars are, in each case, the sets of images under H and G of the vertices of Y. But here we have the following

Lemma 7 *Let* $\Theta : \widetilde{Y} \longrightarrow Y$ *be a locally injective morphism of a connected graph* $\widetilde{Y}$ *into a tree* Y. *Then* Θ *is injective.*

Proof We have to prove that Θ maps distinct vertices of $\widetilde{Y}$ to distinct vertices of Y. To this end, let $\widetilde{P}$ and $\widetilde{Q}$ be distinct vertices of $\widetilde{Y}$ and suppose that they have the same image under Θ. Let $\widetilde{y}_1, \ldots, \widetilde{y}_n$ be a path from $\widetilde{P}$ to $\widetilde{Q}$ without any backtrackings. Then $y_1, \ldots, y_n$ is a closed path in Y, where here $y_i = \Theta(\widetilde{y}_i)$. Since Y is a tree, there must be a backtracking, say

$$y_{i+1} = \overline{y_i}.$$

Now in $\widetilde{Y}$ the star of the terminus of $\widetilde{y}_i$ contains the distinct vertices $o(\widetilde{y}_i)$ and $t(\widetilde{y_{i+1}})$. Since Θ is locally injective, it must map these vertices to distinct vertices in Y. This is a contradiction, and so completes the proof of Lemma 7.

We are now in a position to complete the proof of Theorem 2. Indeed, we claim that the mapping ϕ is an isomorphism. In fact, suppose that $a \in \mathrm{Ker}\phi$, $a \neq 1$. If $\widetilde{P} \in V(\widetilde{Y})$, it follows that $a \notin G_P$ and hence that

$$\widetilde{P} \neq a\widetilde{P}.$$

But

$$\Theta\widetilde{P} = \Theta(a\widetilde{P}),$$

a contradiction.

10 Some consequences of Theorems 2 and 3

Let $G = \pi_1(\mathcal{G}, Y)$ be the fundamental group of a graph of groups $(\mathcal{G}, Y)$. Then G acts on a tree $\widetilde{Y}$. So if $H \leq G$, H acts also on $\widetilde{Y}$. Hence

$$H = \pi_1(\mathcal{H}, X)$$

is again a fundamental group of a graph $(\mathcal{H}, X)$ of groups. This simple observation embodies a rather remarkable subgroup theorem, as one sees from the following applications of Theorem 2. The first of these requires the following definition.

Definition 18 *A group G is said to act freely on a tree if it acts without inversion and only the identity element fixes a vertex.*

Theorem 4 *Let G act freely on a tree. Then G is free.*

Proof As usual

$$G \cong \pi_1(\mathcal{G}, Y, T).$$

So

$$G = \langle\, G_\tau\,,\, t_y\ (y \in E_+(Y) - E(T))\,;\, t_y\, a^y\, t_y^{-1} = a^{\overline{y}}\ (a \in G_y, y \in E_+(Y) - E(T))\,\rangle$$

is an HNN extension with base G_τ. But G_τ is generated by the stabilizers of the vertices of a lift of T. Since G acts freely, these stabilizers are all trivial. Hence $G_\tau = 1$ and

$$G = \langle\, t_y\ (y \in E_+(Y) - E(T))\,\rangle$$

is free on the t_y. ∎

We have already noted that the Cayley graph of a free group, relative to a free set of generators, is a tree. And the free group then acts freely on this tree. Hence so does every one of its subgroups. It follows then from Theorem 4 that subgroups of free groups are free.

For our next examples we need some extra information.

Lemma 8 *Let G be a group acting on a set X. Then*

$$G_{gx} = g G_x g^{-1} \quad (g \in G, x \in X) .$$

Here G_z denotes the stabilizer of $z \in X$.

The proof is easy and is left to the reader.

Lemma 9 *Let G be a group, $A \leq G$ and let*

$$X = \{ gA \mid g \in G \}$$

be the set of all left cosets of A in G. Let H now be a second subgroup of G. If we let H act on X by left multiplication, then

$$H_{gA} = g A g^{-1} \cap H .$$

The proof is straightforward and is left to the reader.

We come now to our second illustration.

Theorem 5 (A.G. Kurosh) *Let*

$$G = A * B$$

be the free product of its subgroups A and B and let $H \leq G$. Then H is a free product of conjugates of subgroups of A and B and a free group.

Proof We note first that G acts on a tree X. Recall that
(1) $V(X) = \{ gA \mid g \in G \} \cup \{ gB \mid g \in G \}$.
(2) $E_+(X) = \{ g \mid g \in G \}$.
So a positive edge takes the form

$$\begin{array}{ccc} & g & \\ \bullet & \longrightarrow & \bullet \\ gA & & gB \end{array}$$

Now H acts also on X which implies that

$$H = \pi_1(\mathcal{H}, Y, T),$$

where

$$Y = H\backslash X$$

and T is a maximal tree in Y. So

$$H = \langle\, H_T\,,\, t_y\, (y \in E_+(Y) - E(T))\,;\, t_y\, a^y\, t_y^{-1} = a^{\overline{y}}\, (a \in H_y\,,\, y \in E_+(Y) - E(T))\,\rangle\,.$$

These edge groups H_y are easy to determine. They are simply of the form

$$H_y = H_{qy} = H_g = \{\, h \mid hg = g\,\} = 1\,,$$

i.e., the edge groups are trivial.

Next we need to know the vertex groups H_P $(P \in V(Y))$. But as before we find that either

$$H_P = H_{qP} = H_{gA} = gAg^{-1} \cap H\,,$$

or

$$H_P = H_{qP} = H_{gB} = gBg^{-1} \cap H\,.$$

So H_T is a free product of conjugates of subgroups of A and B (since the edge groups are all trivial) and hence H is a free product of the free group on the t_y and H_T, as claimed.

Similarly we can deduce

Theorem 6 (Hanna Neumann, A. Karrass and D. Solitar) *Suppose that*

$$G = A \underset{C}{*} B$$

is an amalgamated product. Then every subgroup H of G is an HNN extension of a tree product (i.e., an H_T, where T is a maximal tree in a graph Y) in which the vertex groups are conjugates of subgroups of either A or B, and the edge groups are conjugates of subgroups of C. The associated subgroups are conjugates of subgroups of C.

We need only recall that G acts on a tree X whose edges are all of the form

$$\begin{array}{ccc} & gC & \\ \bullet & \xrightarrow{\quad\quad} & \bullet \\ gA & & gB \end{array}$$

One can then deduce the following

Theorem 7 (Karrass and Solitar) *Let*

$$G = A \underset{C}{*} B$$

where A and B are free and C is cyclic. Then every finitely generated subgroup of G is finitely presented.

The proof of Theorem 7 requires a number of lemmas, which we record here without giving detailed proofs. These we leave to the reader.

Lemma 10 *Let G be a group and suppose that*

$$G = \bigcup_{n=1}^{\infty} G_n.$$

If

$$G_1 < G_2 < \dots$$

is a properly ascending series of subgroups of G, then G is not finitely generated.

Lemma 11 *Let*
$$G = A \underset{C}{*} B.$$

If $A_1 \leq A$, $B_1 \leq B$ and
$$A_1 \cap C = C_1 = B_1 \cap C,$$

then
$$\mathrm{gp}(A_1, B_1) = A_1 \underset{C_1}{*} B_1.$$

The following corollary is an easy consequence of Lemma 11.

Corollary 1 *If C is finitely generated and A is not, then G is not finitely generated.*

Lemma 12 *Let*

$$E = < B, t_1, \dots, t_h; t_i^{-1} L_i t_i = K_i \ (i = 1, \dots, h) >$$

be an HNN extension. Suppose that C is a subgroup of B and that

$$t_i^{-1}(C \cap L_i)t_i = C \cap K_i \ (i = 1, \dots, h).$$

Then

$$\mathrm{gp}(C, t_1, \dots, t_h) = < C, t_1, \dots, t_h; t_i^{-1}(C \cap L_i)t_i = C \cap K_i \ (i = 1, \dots, h) >$$

is again an HNN extension, as indicated.

The following corollary is an easy consequence of Lemma 12.

Corollary 1 *Suppose that*

$$E = <B, t_1, \ldots, t_h; t_i^{-1} L_i t_i = K_i (i = 1, \ldots, h),$$

where $K_1, \ldots, K_k$ are all finitely generated. If B is not finitely generated, then neither is E.

Lemma 13 *Let $(\mathcal{G}, T)$ be a graph of groups, where T is a tree and let*

$$\pi_1(\mathcal{G}, T, T)$$

be the fundamental group of this graph of groups, at the tree T. Then G is finitely generated, if and only if, there exists a finite subtree U of T such that

$$G = \pi_1(\mathcal{G}, U, U)$$

and all the vertex groups of $(\mathcal{G}, U)$ are finitely generated.

11 The tree of SL₂

Our next objective is to show that if F is a field with a discrete valuation then there is a tree X upon which $SL_2(F)$ acts. Let me begin by recalling some definitions and facts.

Definition 19 *Let F be a commutative field, $F^* = F - \{0\}$ viewed as a multiplicative group. Then a surjective map*

$$v : F \longrightarrow \mathbf{Z} \cup \{\infty\}$$

is called a discrete valuation if
(i) $v(0) = \infty$ where $a + \infty = \infty = \infty + a$ for every $a \in F$;
(ii) $v : F^* \longrightarrow \mathbf{Z}$ is a surjective homomorphism from the multiplicative group F^* to the additive group $\mathbf{Z}$, i.e.,

$$v(xy) = v(x) + v(y) \qquad (x, y \in F^*) \; ;$$

(iii) $v(x + y) \geq \min\{v(x), v(y)\}$, where $\infty \geq a$ for every $a \in F$.

We form
$$\mathcal{O} = \{ x \in F \mid v(x) \geq 0 \}$$

the *valuation ring* of F. It is not hard to see that $\mathcal{O}$ is a subring of F. It has the additional property that if $a \in F$, $a \neq 0$ then either $a \in \mathcal{O}$ or $a^{-1} \in \mathcal{O}$.

We gather together some properties of $\mathcal{O}$.

Lemma 14 (i) *The non-units of $\mathcal{O}$ form a maximal ideal $\mathcal{M}$.*
(ii) *$\mathcal{O}$ is a principal ideal domain.*

Proof (i) Notice first that if $a \in \mathcal{O}$ and $v(a) = 0$, then $v(a^{-1}) = -v(a) = 0$. So $a^{-1} \in \mathcal{O}$. Thus the set $\mathcal{M}$ of non-units of $\mathcal{O}$ consists of the elements $a \in \mathcal{O}$ with $v(a) > 0$. This is clearly an ideal by the properties of v. Moreover $\mathcal{M}$ is maximal since the elements outside $\mathcal{M}$ are invertible, i.e., $\mathcal{O}/\mathcal{M}$ is a field.

(ii) Let $\mathcal{I} \neq \mathcal{O}$ be a non-zero ideal of $\mathcal{O}$ and let $\pi \in \mathcal{O}$ be chosen so that

$$v(\pi) = 1 \ .$$

(π is sometimes called a *uniformizer*.) Now if

$$m = \min \left\{ v(a) \mid a \in \mathcal{I} \right\}$$

we claim that $\mathcal{I}$ is the ideal (π^m) generated by π^m. To prove this, first note that if $a \in \mathcal{I}$, $v(a) = m$ then

$$v(a\pi^{-m}) = 0 \ .$$

Hence $a\pi^{-m} \in \mathcal{O}$ and is invertible in $\mathcal{O}$; thus

$$\pi^m = au \in \mathcal{I} \qquad (u \text{ a unit in } \mathcal{O}).$$

So

$$a \in (\pi^m) \ .$$

That the other elements of $\mathcal{I}$ are also contained in (π^m) requires the use of the Euclidean algorithm and uses the minimality of m.

Incidentally it follows from this argument that π generates $\mathcal{M}$.

Now suppose that V is a 2-dimensional left vector space over F. We can also think of V as a left $\mathcal{O}$-module.

Definition 20 *An $\mathcal{O}$-lattice L of V is any $\mathcal{O}$-submodule of the form*

$$L = \mathcal{O}x + \mathcal{O}y \qquad (xy \in V)$$

where x and y are linearly independent over F.

The group F* acts on the set of all such $\mathcal{O}$-lattices L of V by left multiplication:

$$aL = \mathcal{O}ax + \mathcal{O}ay \ .$$

The orbit of such a lattice L under this action is called its *class* and will be denoted by cl(L). Two $\mathcal{O}$-lattices are termed *equivalent* if they lie in the same class. We denote by X the set of all these classes of $\mathcal{O}$-lattices. Our claim is that X can be thought of as a tree. This will follow in due course.

Suppose that L and L′ are $\mathcal{O}$-lattices in V. Then $(L + L')/L'$ is a finitely generated torsion $\mathcal{O}$-module. A submodule of a free module over a principal ideal domain is again free. This leads to the conclusion that a finitely generated module over a principal ideal domain is a direct sum of cyclic modules. It follows that

$$\left((L + L')/L' \cong\right) \ L'/(L \cap L') \cong \mathcal{O}/\pi^c\mathcal{O} \oplus \mathcal{O}/\pi^d\mathcal{O} \qquad (c, d \geq 0) \ .$$

Consequently

$$\pi^e L' \ \leq \ L \quad \text{for some} \quad e \geq 0.$$

Now again the "basis theorem" for submodules of finitely generated free modules over principal ideal domains allows us to choose a basis x, y for L such that

$$\left\{ \pi^f x, \ \pi^g y \right\} \quad \text{is a basis for} \quad \pi^e L'.$$

Hence

$$L' \quad \text{has a basis} \quad \left\{ \pi^i x, \ \pi^j y \right\}.$$

We claim that

$$|i - j| = |c - d|$$

depends only on cl(L) and cl(L′). In order to prove this statement consider instead the $\mathcal{O}$-lattices aL and bL' ($a, b \in F^*$). Then aL has a basis $\{ax, ay\}$ and bL' a basis $\{b\pi^i x, b\pi^j y\}$. Notice that if $v(ba^{-1}) = n$, then $ba^{-1} = u\pi^n$ (u a unit in $\mathcal{O}$). So

$$b = au\pi^n \ .$$

Hence bL' has an $\mathcal{O}$-basis $\left\{ \pi^{i+n}(ax), \ \pi^{j+n}(ay) \right\}$ and once again

$$|(i + n) - (j + n)| = |c - d|$$

as claimed.

If we now denote cl(L) by Λ and cl(L′) by Λ' then we *define*

$$d(\Lambda, \Lambda') = |c - d|$$

and term $d(\Lambda, \Lambda')$ the *distance between Λ and Λ'*. We term Λ and Λ' *adjacent* if

$$d(\Lambda, \Lambda') = 1 \ .$$

This allows us to turn X, the set of all such equivalence classes Λ of $\mathcal{O}$-lattices in V, into a graph. Here

$$V(X) = \left\{\ \Lambda \ \mid\ \Lambda = \mathrm{cl}(L),\ L \text{ an } \mathcal{O} - \text{lattice in } V\ \right\}\ .$$

We then define

$$E(X) = \left\{\ (\Lambda, \Lambda')\ \mid\ \Lambda, \Lambda' \text{ adjacent}\ \right\}\ .$$

For each such edge (Λ, Λ') we define

$$o\big((\Lambda, \Lambda')\big) = \Lambda\quad ,\quad t\big((\Lambda, \Lambda')\big) = \Lambda'\ .$$

Finally we define

$$\overline{(\Lambda, \Lambda')} = (\Lambda', \Lambda)\ .$$

What we have left then is to prove the

Theorem 8 X *is a tree.*

The proof is not hard, once one figures out what has to be proved. It can be found on page 70 of Serre's book cited at the beginning of this chapter

Now $GL(V)$ acts on X in the obvious way and so $SL(V)$ does as well. Unfortunately $GL(V)$ acts on X with inversion, but $SL(V)$ acts without inversion. So our structure theorems apply to $SL(V)$. Indeed one can, e.g., deduce

Theorem 9 (Ihara) $SL_2(F)$ *is an amalgamated product:*

$$SL_2(F) = SL_2(\mathcal{O}) \underset{\Gamma}{*} SL_2(\mathcal{O})$$

where here Γ is the subgroup of $SL_2(\mathcal{O})$ consisting of the matrices

$$\begin{pmatrix} a & b \\ c & d \end{pmatrix}\qquad c \equiv 0\,(\mathrm{mod}\,\pi)\ .$$

Now suppose that G is a group and that there exist representations of G in $SL(V)$. Such representations provide us with an action of G on the tree X of $SL(V)$ that we discussed above and as a consequence yields a description of G as the fundamental group of a graph of groups:

$$G \cong \pi_1(\mathcal{G}, Y)\ .$$

Unravelling stabilizers of edges and vertices is the next task if this description is to be useful. In the event that one is able to find the right kind of representations of G, the task of geometric representation theory, then this approach does turn out to be useful. This point of view was introduced by Culler and Shalen in their fundamental paper on the fundamental groups of three manifolds. Similar techniques apply also to groups given by generators and defining relations and yield, in particular, a proof of Theorem 8 of Chapter V.

Index

Lectures in Mathematics - ETH Zürich

Each year the Eidgenössische Technische Hochschule (ETH) at Zürich invites a selected group of mathematicians to give postgraduate seminars in various areas of pure and applied mathematics. These seminars are directed to an audience of many levels and backgrounds. Now some of the most successful lectures are being published for a wider audience through the **Lectures in Mathematics - ETH Zürich** *series. Lively and informal in style, moderate in size and price, these books will appeal to professionals and students alike, bringing a quick understanding of some important areas of current research.*

Previously published:

Randall J. LeVeque, Numerical Methods for Conservation Laws.
Second edition 1992, 214 pages, softcover, ISBN 3-7643-2723-5.

J. Donald Monk, Cardinal Functions on Boolean Algebras.
1990, 152 pages, softcover, ISBN 3-7643-2495-3.

Carl de Boor, Splinefunktionen (german).
1990, 184 pages, softcover, ISBN 3-7643-2514-3.

Daniel Bättig/Horst Knörrer, Singularitäten (german).
1991, 140 pages, softcover, ISBN 3-7643-2616-6.

Anthony J. Tromba, Teichmüller Theory in Riemannian Geometry.
1992, 224 pages, softcover, ISBN 3-7643-2735-9.

Raghavan Narasimhan, Compact Riemann Surfaces.
1992, 128 pages, softcover, ISBN 3-7643-2742-1.

Marc Yor, Some Aspects of Brownian Motion. Part I: Some Special Functionals.
1992, 144 pages, softcover, ISBN 3-7643-2807-X.

Olavi Nevanlinna, Convergence of Iterations for Linear Equations.
1993, 184 pages, softcover, ISBN 3-7643-2865-7.

Gilbert Baumslag, Topics in Combinatorial Group Theory.
1993, 172 pages, softcover, ISBN 3-7643-2921-1.

Mariano Giaquinta, Introduction to Regularity Theory for Nonlinear Elliptic Systems.
1993, 144 pages, softcover, ISBN 3-7643-2879-7.

Monographs in Mathematics

Managing Editors:

H. Amann (Universität Zürich)
K. Grove (University of Maryland, College Park)
H. Kraft (Universität Basel)
P.-L. Lions (Université de Paris-Dauphine)

Associate Editors:

H. Araki (Kyoto University)
J. Ball (Heriot-Watt University, Edinburgh)
F. Brezzi (Università di Pavia)
K.C. Chang (Peking University)
N. Hitchin (University of Warwick)
H. Hofer (Universität Bochum)
H. Knörrer (ETH Zürich)
K. Masuda (University of Tokyo)
D. Zagier (Max-Planck-Institut, Bonn)

The foundations of this outstanding book series were laid in 1944. Until the end of the 1970s, a total of 77 volumes appeared, including works of such distinguished mathematicians as Carathéodory, Nevanlinna, and Shafarevich, to name a few. The series came to its present name and appearance in the 1980s. According to its well-established tradition, only monographs of excellent quality will be published in this collection. Comprehensive, in-depth treatments of areas of current interest are presented to a readership ranging from graduate students to professional mathematicians. Concrete examples and applications both within and beyond the immediate domain of mathematics illustrate the import and consequences of the theory under discussion.

We encourage preparation of manuscripts in TeX for delivery in camera-ready copy which leads to rapid publication, or in electronic form for interfacing with laser printers or typesetters. Proposals should be sent directly to the editors or to: Birkhäuser Verlag, P.O. Box 133, CH-4010 Basel, Switzerland.